Christian Dawson (Ed.)

Applied Artificial Neural Networks

MDPI

This book is a reprint of the Special Issue that appeared in the online, open access journal, *Applied Sciences* (ISSN 2076-3417) from 2015–2016, available at:

http://www.mdpi.com/journal/applsci/special_issues/neural_network

Guest Editor
Christian Dawson
Computer Science Department, Loughborough University
UK

Editorial Office
MDPI AG
St. Alban-Anlage 66
Basel, Switzerland

Publisher
Shu-Kun Lin

Senior Assistant Editor
Yurong Zhang

1. Edition 2016

MDPI • Basel • Beijing • Wuhan • Barcelona • Belgrade

ISBN 978-3-03842-270-9 (Hbk)
ISBN 978-3-03842-271-6 (PDF)

Table of Contents

V

List of Contributors

Ashfaq Ahmad COMSATS Institute of Information Technology, Islamabad 44000, Pakistan.

Roberto Alejo Pattern Recognition Laboratory, Tecnológico de Estudios Superiores de Jocotitlán, Carretera Toluca-Atlacomulco KM 44.8, Ejido de San Juan y San Agustín, Jocotitlán 50700, Mexico.

Nabil Alrajeh College of Applied Medical Sciences, Department of Biomedical Technology, King Saud University, Riyadh 11633, Saudi Arabia.

Juan A. Antonio-Velázquez Pattern Recognition Laboratory, Tecnológico de Estudios Superiores de Jocotitlán, Carretera Toluca-Atlacomulco KM 44.8, Ejido de San Juan y San Agustín, Jocotitlán 50700, Mexico.

Mao-Yong Cao College of Electronics, Communication and Physics, Shandong University of Science and Technology, Qingdao 266590, China.

Fa-Liang Chang School of Control Science and Engineering, Shandong University, Jinan 250061, China.

Pan-Fei Chen College of Electronics, Communication and Physics, Shandong University of Science and Technology, Qingdao 266590, China.

Zhenguo Chen Computer Department, North China Institute of Science and Technology, East Yanjiao, Beijing 101601, China.

Kewei Cheng School of Computing, Informatics, Decision Systems Engineering (CIDSE), Ira A. Fulton Schools of Engineering, Arizona State University, Tempe 85281, AZ, USA.

Zhengchao Dong Translational Imaging Division, Columbia University, New York, NY 10032, USA; State Key Lab of CAD & CG, Zhejiang University, Hangzhou 310027, China.

Zong Woo Geem Department of Energy and Information Technology, Gachon University, 1342 Seongnamdae-ro, Sujeong-gu, Seongnam-si, Gyeonggi-do 13120, Korea.

Wanfei He Department of Art, Jincheng College of Sichuan University, Chengdu 610000, Sichuan, China.

Yu-Han Hou College of Electronics, Communication and Physics, Shandong University of Science and Technology, Qingdao 266590, China.

Nadeem Javaid COMSATS Institute of Information Technology, Islamabad 44000, Pakistan.

Abid Khan COMSATS Institute of Information Technology, Islamabad 44000, Pakistan.

Zahoor Ali Khan Internetworking Program, Faculty of Engineering, Dalhousie University, Halifax, NS, B3J 4R2, Canada; Computer Information Science, Higher Colleges of Technology, Fujairah Campus 4114, Abu Dhabi 17666, United Arab Emirates.

Anzy Lee Department of Civil and Environmental Engineering, Seoul National University, 1 Gwanak-ro, Gwanak-gu, Seoul 08826, Korea.

Chengguang Li College of Computer Science and Engineer, Northeastern University, Shenyang 110819, China.

Fangfang Li College of Water Resources & Civil Engineering, China Agricultural University, Beijing 100083, China.

Hao Li College of Chemistry, Sichuan University, Chengdu 610064, China.

Xueying Li State Key Laboratory of Simulation and Regulation of Water Cycle in River Basin, China Institute of Water Resources and Hydropower Research, Beijing 100038, China; College of Water Resources & Civil Engineering, China Agricultural University, Beijing 100083, China.

Zijun Li College of Light Industry, Textile and Food Engineering, Sichuan University, Chengdu 610065, Sichuan, China.

Fan Lin Software School, Xiamen University, Xiamen 361005, China.

Wei-Jian Liu College of Electronics, Communication and Physics, Shandong University of Science and Technology, Qingdao 266590, China.

Zhijian Liu Department of Power Engineering, School of Energy, Power and Mechanical Engineering, North China Electric Power University, Baoding 071003, China.

Erika López-González Pattern Recognition Laboratory, Tecnológico de Estudios Superiores de Jocotitlán, Carretera Toluca-Atlacomulco KM 44.8, Ejido de San Juan y San Agustín, Jocotitlán 50700, Mexico.

Siyuan Lu Jiangsu Key Laboratory of 3D Printing Equipment and Manufacturing, Nanjing 210042, China; Key Laboratory of Symbolic Computation and Knowledge Engineering of Ministry of Education, Jilin University, Changchun 130012, China.

Juan Monroy-de-Jesús Computer Science, Universidad Autónoma del Estado de México, Carretera Toluca- Atlacomulco KM 60, Atlacomulco 50000, Mexico.

Juan H. Pacheco-Sánchez Division of Graduate Studies and Research, Instituto Tecnológico de Toluca, Av. Tecnológico s/n. Colonia Agrícola Bellavista, Metepec, Edo. De México 52149, Mexico.

Jia Pan State Key Laboratory of Mechanical Transmission, School of Material Science and Engineering, Chongqing University, Chongqing 400044, China.

Zhen Peng Information Management Department, Beijing Institute of Petrochemical Technology, Beijing 100029, China.

Umar Qasim Cameron Library, University of Alberta, Edmonton, AB, T6G 2J8 Canada.

Jun Qiu Institute for Aero-Engine, School of Aerospace Engineering, Tsinghua University, Beijing 100084, China.

Guo-zheng Quan State Key Laboratory of Mechanical Transmission, School of Material Science and Engineering, Chongqing University, Chongqing 400044, China.

Rong Shan College of Electronics, Communication and Physics, Shandong University of Science and Technology, Qingdao 266590, China.

Qianqian Shang Nanjing Hydraulic Research Institute, Nanjing 210029, China.

Jiadong Shi School of Mechatronical Engineering, Beijing Institute of Technology, 5 South Zhongguancun Street, Haidian District, Beijing 100081, China.

Kyung-Duck Suh Department of Civil and Environmental Engineering, Seoul National University, 1 Gwanak-ro, Gwanak-gu, Seoul 08826, Korea.

Xindong Tang College of Mathematics, Sichuan University, Chengdu 610064, China.

Jianzhong Wang School of Mechatronical Engineering, Beijing Institute of Technology, 5 South Zhongguancun Street, Haidian District, Beijing 100081, China.

Run Wang College of Light Industry, Textile and Food Science Engineering, Sichuan University, Chengdu 610064, China.

Shuihua Wang School of Computer Science and Technology & School of Psychology, Nanjing Normal University, Nanjing 210023, China; Key Laboratory of Statistical information Technology and Data Mining, State Statistics Bureau, Chengdu 610225, China.

Xuan Wang State Key Laboratory of Mechanical Transmission, School of Material Science and Engineering, Chongqing University, Chongqing 400044, China.

Zhigang Wang Key Laboratory of Computer Vision and System, Ministry of Education, Tianjin Key Laboratory of Intelligence Computing and Novel Software Technology, Tianjin University of Technology, Tianjin 300384, China.

Lifeng Wu College of Information Engineering, Capital Normal University, Beijing 100048, China.

Shu-Yi Xiao College of Electronics, Communication and Physics, Shandong University of Science and Technology, Qingdao 266590, China.

Jiquan Yang Jiangsu Key Laboratory of 3D Printing Equipment and Manufacturing, Nanjing 210042, China.

Ming Yang Department of Radiology, Nanjing Children's Hospital, Nanjing Medical University, Nanjing 210008, China.

Ying Yin College of Computer Science and Engineer, Northeastern University, Shenyang 110819, China.

Bin Zhang College of Computer Science and Engineer, Northeastern University, Shenyang 110819, China.

Guangyue Zhang School of Mechatronical Engineering, Beijing Institute of Technology, 5 South Zhongguancun Street, Haidian District, Beijing 100081, China.

Yudong Zhang School of Computer Science and Technology & School of Psychology, Nanjing Normal University, Nanjing 210023, China; Department of Neurology, First Affiliated Hospital of Nanjing Medical University, Nanjing 210029, China; Guangxi Key Laboratory of Manufacturing System & Advanced Manufacturing Technology, Guilin 541004, China.

Min Zhao College of Light Industry, Textile and Food Engineering, Sichuan University, Chengdu 610065, Sichuan, China.

Yuhai Zhao College of Computer Science and Engineer, Northeastern University, Shenyang 110819, China; Key Laboratory of Computer Network and Information Integration, Southeast University, Ministry of Education, Nanjing 211189, China.

Zeng-Shun Zhao College of Electronics, Communication and Physics, Shandong University of Science and Technology, Qingdao 266590, China; School of Control Science and Engineering, Shandong University, Jinan 250061, China.

About the Guest Editor

Christian Dawson is a Senior Lecturer in the Department of Computer Science at Loughborough University, U.K. His research interests include software engineering, artificial intelligence and applications in hydroinformatics. He has authored a number of books and has published over 100 articles in these areas. He has been involved in a number of projects with the Environment Agency and Environment Canada, is a member of a number of conference committees, and an Editorial Board Member of *Computational Intelligence and Neuroscience*.

Preface to "Applied Artificial Neural Networks"

Since their re-popularisation in the mid-1980s, artificial neural networks have seen an explosion of research across a diverse spectrum of areas. While an immense amount of research has been undertaken in artificial neural networks themselves—in terms of training, topologies, types, etc.—a similar amount of work has examined their application to a whole host of real-world problems. Such problems are usually difficult to define and hard to solve using conventional techniques. Examples include computer vision, speech recognition, financial applications, medicine, meteorology, robotics, hydrology, etc.

This Special Issue focuses on the second of these two research themes, that of the application of neural networks to a diverse range of fields and problems. It collates contributions concerning neural network applications in areas such as engineering, hydrology and medicine.

Christian Dawson
Guest Editor

Comparative Study on Theoretical and Machine Learning Methods for Acquiring Compressed Liquid Densities of 1,1,1,2,3,3,3-Heptafluoropropane (R227ea) via Song and Mason Equation, Support Vector Machine, and Artificial Neural Networks

Hao Li, Xindong Tang, Run Wang, Fan Lin, Zhijian Liu and Kewei Cheng

Abstract: 1,1,1,2,3,3,3-Heptafluoropropane (R227ea) is a good refrigerant that reduces greenhouse effects and ozone depletion. In practical applications, we usually have to know the compressed liquid densities at different temperatures and pressures. However, the measurement requires a series of complex apparatus and operations, wasting too much manpower and resources. To solve these problems, here, Song and Mason equation, support vector machine (SVM), and artificial neural networks (ANNs) were used to develop theoretical and machine learning models, respectively, in order to predict the compressed liquid densities of R227ea with only the inputs of temperatures and pressures. Results show that compared with the Song and Mason equation, appropriate machine learning models trained with precise experimental samples have better predicted results, with lower root mean square errors (RMSEs) (e.g., the RMSE of the SVM trained with data provided by Fedele *et al.* [1] is 0.11, while the RMSE of the Song and Mason equation is 196.26). Compared to advanced conventional measurements, knowledge-based machine learning models are proved to be more time-saving and user-friendly.

Reprinted from *Appl. Sci.* Cite as: Li, H.; Tang, X.; Wang, R.; Lin, F.; Liu, Z.; Cheng, K. Comparative Study on Theoretical and Machine Learning Methods for Acquiring Compressed Liquid Densities of 1,1,1,2,3,3,3-Heptafluoropropane (R227ea) via Song and Mason Equation, Support Vector Machine, and Artificial Neural Networks. *Appl. Sci.* **2016**, *6*, 25.

1. Introduction

The increasing problems of greenhouse effect and ozone depletion have drawn people's great attentions during the past decades [2–5]. In the field of heating, ventilation, air conditioning, and refrigeration (HVAC and R) [6–8], scientists started to use 1,1,1,2,3,3,3-heptafluoropropane (R227ea) [9–11] as a substitute in order to replace other refrigerants that are harmful to the ozone (like R114, R12, and R12B1), because R227ea has a zero ozone depletion potential (ODP) [12]. Other applications of

R227ea include the production of rigid polyurethane foams and aerosol sprays [11,13]. R227ea has been shown to be crucial in industrial fields and scientific research.

In practical applications, the use of R227ea requires the exact values of the compressed liquid densities under certain values of temperatures and pressures. However, due to the complexity and uncertainty of the density measurement of R227ea, precise values of the density are usually difficult to acquire. To solve this problem, molecular dynamic (MD) simulation methods [14–16] have been used for predicting related thermophysical properties of refrigerants. Nevertheless, these simulation methods have high requirements for computers and require long computational times. Additionally, they need accurate forms of potential energy functions. Motivated by these issues, here, as a typical case study, we aim at finding out alternative modeling methods to help acquire precise values of the densities of R227ea.

Acquiring the density by theoretical conclusion is an alternative approach to replace the MD methods. Equation of state is one of the most popular descriptions of theoretical studies that illustrates the relationship between temperature, pressure, and volume for substances. Based on the recognition that the structure of a liquid is determined primarily by the inner repulsive forces, the Song and Mason equation [17] was developed in the 1990s based on the statistical-mechanics perturbation theories [18,19] and proved to be available in calculating the densities of various refrigerants recently [20]. However, limitations of the theoretical methods are also apparent. Firstly, the calculated results of refrigerants are not precise enough. Secondly, previous studies only discussed the single result with a given temperature and pressure [20], neglecting the overall change regulation of the density with the changes of temperature and pressure. To find out a better approach that can precisely acquire the density values of R227ea, here, we first illustrate the three-dimensional change regulation of the density of R227ea with the changes of temperature and pressure using the Song and Mason equation, and also use novel machine learning techniques [21–23] to predict the densities of R227ea based on three groups of previous experimental data [1,24,25]. To define the best machine learning methods for the prediction of the densities of R227ea, different models should be evaluated respectively, which is a necessary comparison process in environmental science. In this case study, support vector machine (SVM) and artificial neural networks (ANNs) were developed, respectively, in order to find out the best model for density prediction. ANNs are powerful non-linear fitting methods that developed during decades, which have good prediction results in many environmental related fields [26–30]. However, although ANNs usually give effective prediction performances, there is a risk of over-fitting phenomenon [26] if the best number of hidden nodes are not defined, which also indicates that the data size for model training should be large enough. Additionally, the training of

ANNs may require relatively long training times if the numbers of hidden nodes are high or the data size is large. Alternatively, SVM, a new machine learning technique developed during these years, has been proved to be effective in numerical predictions for environmental fields [26,27]. The SVM is usually considered to have better generalization performance, leading to better predicted results in many scientific cases [26]. Furthermore, a proper training of SVM has fewer requirements to the data size, ensuring that it can be used for dealing with many complicated issues. Despite the advantages of ANNs and SVM, for the prediction of compressed liquid density of R227ea, it is hard to define the best models without studies. Therefore, here, ANNs (with different numbers of hidden nodes) and SVM were developed respectively. Comparisons were made among different methodologies in order to find the best models for practical applications.

2. Experimental Section

2.1. Theoretical Equation of State

Based on statistical-mechanical perturbation theories [18,19], Song and Mason [17] developed a theoretical equation of state to analyze convex-molecular fluids, which is shown in Equation (1):

$$\frac{P}{\rho k_B T} = 1 + B_2(T)\rho + \alpha(T)\rho[G(\eta) - 1] \tag{1}$$

where T is the temperature (K), P is the pressure (bar), ρ is the molar density (kg·m^{-3}), k_B is the Boltzmann constant, $B_2(T)$ is the second virial coefficient, $\alpha(T)$ is the contribution of the repulsive forces to the second virial coefficient, $G(\eta)$ is the average pair distribution function at contact for equivalent hard convex bodies [20], η is the packing fraction. To the convex bodies, $G(\eta)$ can be adopted as follows [17,20]:

$$G(\eta) = \frac{1 - \gamma_1\eta + \gamma_2\eta^2}{(1 - \eta)^3} \tag{2}$$

where γ_1 and γ_2 are values to reproduce the precise third and fourth virial coefficients, which can be estimated as [17,20]:

$$\gamma_1 = 3 - \frac{1 + 6\gamma + 3\gamma^2}{1 + 3\gamma} \tag{3}$$

and

$$\gamma_2 = 3 - \frac{2 + 2.64\gamma + 7\gamma^2}{1 + 3\gamma} \tag{4}$$

3

In terms of η, it holds that

$$\eta = \frac{b(T)\rho}{1 + 3\gamma} \tag{5}$$

where b is the van der Waals convolume, which can be shown with α [17,20]:

$$b(T) = \alpha(T) + T\frac{d\alpha(T)}{dT} \tag{6}$$

$B_2(T)$, $\alpha(T)$ and $b(T)$ can be described in with the temperature of normal boiling point (T_{nb}) and the density at normal boiling point (ρ_{nb}) [17,20]:

$$B_2(T)\rho_{nb} = 1.033 - 3.0069(\frac{T_{nb}}{T}) - 10.588(\frac{T_{nb}}{T})^2 + 13.096(\frac{T_{nb}}{T})^3 - 9.8968(\frac{T_{nb}}{T})^4 \tag{7}$$

and

$$\alpha(T)\rho_{nb} = a_1\left\{\exp\left[-c_1(\frac{T}{T_{nb}})\right]\right\} + a_2\left\{1 - \exp\left[-c_2\left(\frac{T}{T_{nb}}\right)^{-0.25}\right]\right\} \tag{8}$$

and

$$b(T)\rho_{nb} = a_1\left[1 - c_1\left(\frac{T}{T_{nb}}\right)\right]\exp\left[-c_1\left(\frac{T}{T_{nb}}\right)\right] + a_2\left\{1 - \left[1 + 0.25c_2\left(\frac{T_{nb}}{T}\right)^{0.25}\right]\exp\left[-c_2\left(\frac{T}{T_{nb}}\right)^{-0.25}\right]\right\} \tag{9}$$

where $\alpha_1 = -0.086$, $\alpha_2 = 2.3988$, $c_1 = 0.5624$, and $c_2 = 1.4267$.

Now that we have Equations (1)–(9) above, the last values we should know are γ, T_{nb}, and ρ_{nb}. γ can be obtained from fitting the experimental results, and T_{nb} and ρ_{nb} can be obtained from standard experimental data. According to previous studies, for R227ea, γ is 0.760 [20], T_{nb} is 256.65 K [31] and ρ_{nb} is 1535.0 kg·m^{-3} [31]. Now we can only input the values of T (K) and P (bar) to Equation (1) and the calculated density of R227ea can be acquired.

2.2. Support Vector Machine (SVM)

SVM is a powerful machine learning method based on statistical learning theory. On the basis of the limited information of samples, SVM has an extraordinary ability of optimization for improving generalization. The main principle of SVM is to find the optimal hyperplane, a plane that separates all samples with the maximum margin [32,33]. The plane helps improve the predictive ability of the model and reduce the error which occurs occasionally when predicting and classifying. Figure 1 shows the main structure of a SVM [34,35]. The letter "K" represents kernels [36]. As we can see from Figure 1, it is a small subset extracted from the training data by relevant algorithm that consists of the SVM. For practical applications, choosing appropriate kernels and parameters are important for us to acquire better prediction

4

accuracies. However, there is still no existing standard for scientists to choose these parameters. In most cases, the comparison of experimental results, the experiences from copious calculating, and the use of cross-validation that is available in software packages can help us address this problem [34,37,38].

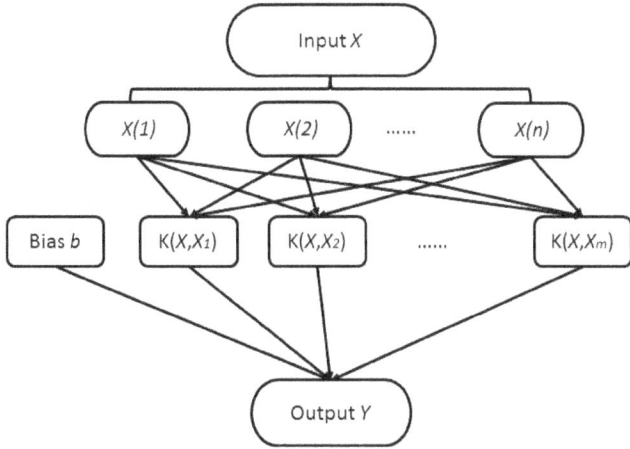

Figure 1. Main structure of a support vector machine (SVM) [35].

2.3. Artificial Neural Networks (ANNs)

ANNs [39–41] are machine learning algorithms with the functions of estimation and approximation based on inputs, which are inspired from the biological neural networks of human brains. Being different from networks with only one or two layers of single direction logic, they use algorithms in control determining and function organizing. The interconnected networks usually consist of neurons that can calculate values from inputs and adapt to different circumstances. Thus, ANNs have powerful capacities in numeric prediction and pattern recognition, which have obtained wide popularity in inferring a function from observation, especially when the object is too complicated to be dealt with by human brains. Figure 2 presents a schematic structure of an ANN for the prediction of compressed liquid density of R227ea, which contains the input layer, hidden layer, and output layer. The input layer consists of two nodes, representing the inputted temperature and pressure, respectively. The output layer is made up of the neuron that represents the density of R227ea.

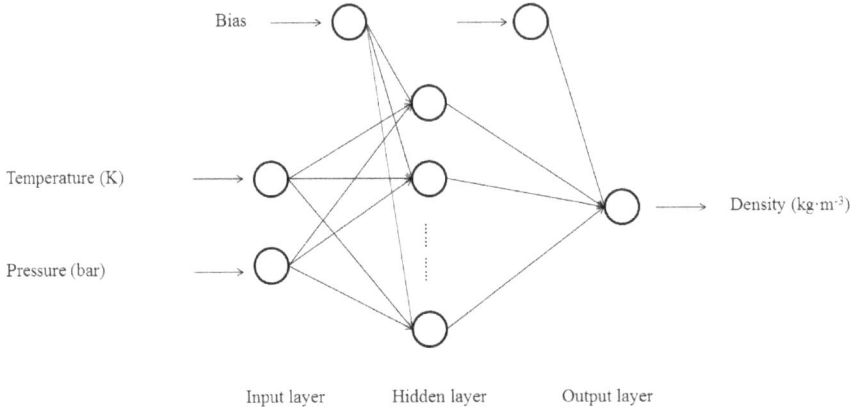

Figure 2. Schematic structure of an artificial neural network (ANN) for the prediction of compressed liquid densities of 1,1,1,2,3,3,3-heptafluoropropane (R227ea).

3. Results and Discussion

3.1. Model Development

3.1.1. Theoretical Model of the Song and Mason Equation

With the Equations (1)–(9) and related constants, the three-dimensional calculated surface of the compressed liquid density of R227ea can be obtained (Figure 3). To make sufficient comparisons between theoretical calculated values and experimental values, previous experimental results provided by Fedele *et al.* (with 300 experimental data groups) [1], Ihmels *et al.* (with 261 experimental data groups) [24], and Klomfar *et al.* (with 83 experimental data groups) [25], were used for making comparisons in Figure 3. It can be seen that though the experimental data is close to the calculated theoretical surface, the theoretical surface does not highly coincide with all the experimental data. We can see that experimental results provided by Fedele *et al.* [1] and Ihmels *et al.* [24] are generally higher than the calculated surface, while the experimental results provided by Klomfar *et al.* [25] have both higher and lower values than the calculated surface. The root mean square errors (RMSEs) of the theoretical calculated results with the three experimental results are 196.26, 372.54, and 158.54, respectively, which are relatively high and not acceptable to practical applications. However, it should be mentioned that the tendency of the surface is in good agreement with the tendency of the experimental data provided by Fedele *et al.* [1] and Ihmels *et al.* [24]. Interestingly, it is obvious to find that when the temperature is close to 100 K, the density would become increasingly high, which has not been reported by experimental results so far.

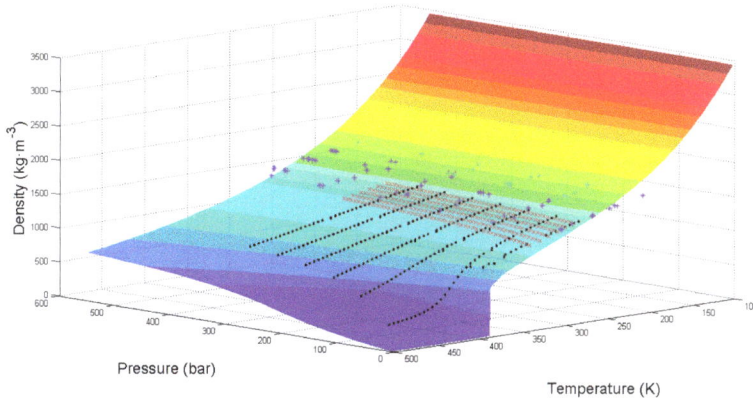

Figure 3. Theoretical calculated surface and experimental densities of R227ea. The surface represents the theoretical calculated results by Equations (1)–(9); black points represent the experimental results from Fedele *et al.* [1]; red crosses represent the experimental results from Ihmels *et al.* [24]; blue asterisks represent the experimental results from Klomfar *et al.* [25].

3.1.2. Machine Learning Models

To develop predictive models via machine learning, we should first define the independent variables and the dependent variable. With the experimental fact during the practical measurements, the temperature and pressure of R227ea are easy to obtain. Here, we define the temperature (K) and pressure (bar) of the determinant as the independent variables, while the density (kg·m^{-3}) is set as the dependent variable. With the design that users can only input the values of the temperature and pressure to a developed model, we let the machine learning models in our study "learn" the existing data and make precise predictions. The experimental data of Fedele *et al.* [1], Ihmels *et al.* [24], and Klomfar *et al.* [25] were used for model developments respectively. In each model, 80% of the data were set as the training set, while 20% of the data were set as the testing set. The SVMs were developed by Matlab software (Libsvm package [42]) and the ANNs were developed by NeuralTools® software (trial version, Palisade Corporation, NY, USA). General regression neural network (GRNN) [43–45] and multilayer feed-forward neural networks (MLFNs) [46–48] were chosen as the learning algorithms of ANNs. Numbers of nodes in the hidden layer of MLFNs were set from 2 to 35. In this case study, the number of hidden layer was set as one. Trials of all ANNs were set as 10,000. All these settings of ANNs were set directly in the NeuralTools® software. Linear regression models were also developed for comparisons. To measure the performance of the model and make suitable comparisons, RMSE (for testing), training time, and prediction accuracy (under the tolerance of 30%) were used as indicators that evaluate the

7

models. Model results using experimental data from Fedele *et al.* [1], Ihmels *et al.* [24], and Klomfar *et al.* [25] are shown in Tables 1–4 respectively. Error analysis results are shown in Figure 4.

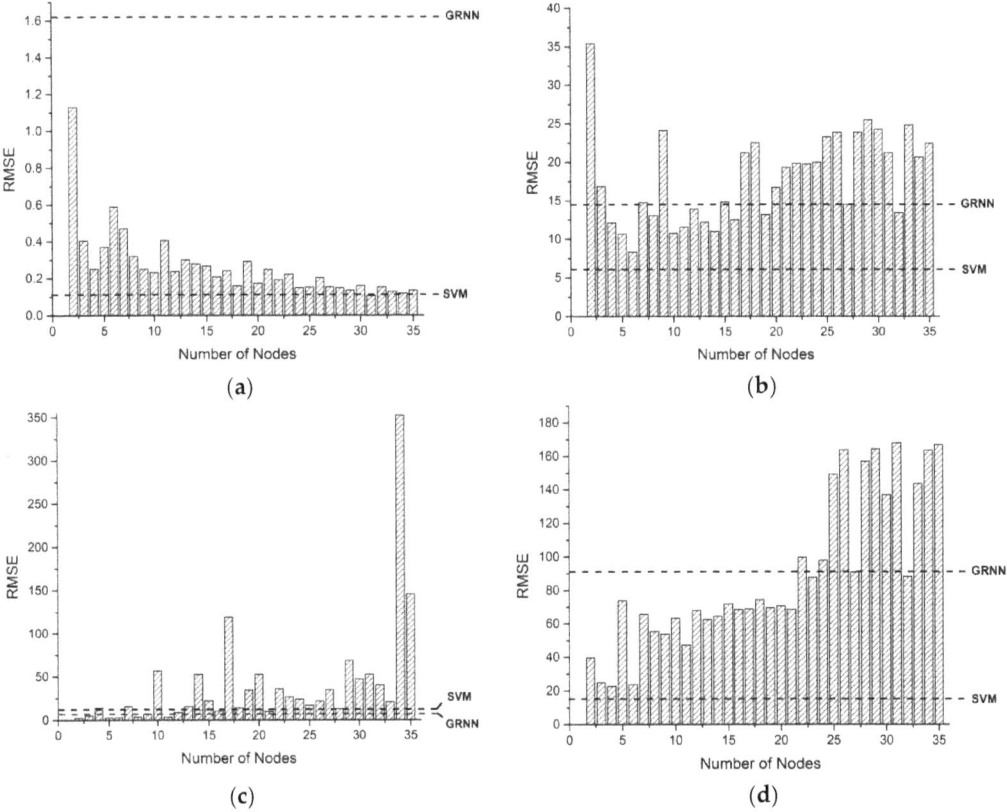

Figure 4. Root mean square error (RMSE) *versus* number of nodes of multilayer feed-forward neural networks (MLFNs). Bars represent the RMSEs; black dashed lines represent the RMSEs of general regression neural network (GRNN) and support vector machine (SVM). (**a**) Machine learning models for data provided by Fedele *et al.* [1]; (**b**) machine learning models for data provided by Ihmels *et al.* [24]; (**c**) machine learning models for data provided by Klomfar *et al.* [25]; and (**d**) machine learning models for data provided by all the three experimental reports [1,24,25].

Table 1. Prediction models using experimental data by Fedele *et al.* [1].

Model Type	RMSE (for Testing)	Training Time	Prediction Accuracy
Linear Regression	10.90	0:00:01	85.0%
SVM	0.11	0:00:01	100%
GRNN	1.62	0:00:01	100%
MLFN 2 Nodes	1.13	0:03:46	100%
MLFN 3 Nodes	0.40	0:04:52	100%
MLFN 4 Nodes	0.25	0:06:33	100%
MLFN 5 Nodes	0.37	0:07:25	100%
MLFN 6 Nodes	0.59	0:10:38	100%
MLFN 7 Nodes	0.47	0:13:14	100%
MLFN 8 Nodes	0.32	0:14:10	100%
. . .	. . .	. . .	. . .
MLFN 29 Nodes	0.13	2:00:00	100%
MLFN 30 Nodes	0.16	2:00:00	100%
MLFN 31 Nodes	0.10	2:00:00	100%
MLFN 32 Nodes	0.15	2:00:00	100%
MLFN 33 Nodes	0.13	2:00:00	100%
MLFN 34 Nodes	0.12	2:00:00	100%
MLFN 35 Nodes	0.13	2:00:00	100%

Root mean square error (RMSE); Support vector machine (SVM); General regression neural network (GRNN); Multilayer feed-forward neural network (MLFN).

Table 2. Prediction models using experimental data by Ihmels *et al.* [24].

Model Type	RMSE (for Testing)	Training Time	Prediction Accuracy
Linear Regression	86.33	0:00:01	63.4%
SVM	6.09	0:00:01	100%
GRNN	14.77	0:00:02	96.2%
MLFN 2 Nodes	35.41	0:02:18	82.7%
MLFN 3 Nodes	16.84	0:02:55	96.2%
MLFN 4 Nodes	12.14	0:03:38	96.2%
MLFN 5 Nodes	10.67	0:04:33	96.2%
MLFN 6 Nodes	8.35	0:04:54	98.1%
MLFN 7 Nodes	14.77	0:06:06	96.2%
MLFN 8 Nodes	13.06	3:19:52	96.2%
. . .	. . .	. . .	. . .
MLFN 29 Nodes	25.46	0:31:00	90.4%
MLFN 30 Nodes	24.25	0:34:31	90.4%
MLFN 31 Nodes	21.23	0:42:16	90.4%
MLFN 32 Nodes	13.40	3:38:17	96.2%
MLFN 33 Nodes	24.84	0:47:06	90.4%
MLFN 34 Nodes	20.65	0:53:14	90.4%
MLFN 35 Nodes	22.46	0:58:16	90.4%

Table 3. Prediction models using experimental data by Klomfar *et al.* [25].

Model Type	RMSE (for Testing)	Training Time	Prediction Accuracy
Linear Regression	15.87	0:00:01	94.1%
SVM	13.93	0:00:01	94.1%
GRNN	9.53	0:00:01	100%
MLFN 2 Nodes	2.72	0:01:13	100%
MLFN 3 Nodes	5.10	0:01:19	100%
MLFN 4 Nodes	14.05	0:01:36	94.1%
MLFN 5 Nodes	2.77	0:02:25	100%
MLFN 6 Nodes	2.85	0:02:31	100%
MLFN 7 Nodes	15.72	0:03:15	94.1%
MLFN 8 Nodes	3.46	0:03:40	100%
. . .	. . .	. . .	. . .
MLFN 29 Nodes	68.34	0:15:03	82.4%
MLFN 30 Nodes	47.09	0:17:58	82.4%
MLFN 31 Nodes	52.60	0:22:01	82.4%
MLFN 32 Nodes	40.03	0:27:46	82.4%
MLFN 33 Nodes	20.69	0:39:27	94.1%
MLFN 34 Nodes	352.01	0:56:26	11.8%
MLFN 35 Nodes	145.61	5:01:57	11.8%

Table 1 and Figure 4a show that the prediction results of machine learning models are generally acceptable, with lower RMSEs than that of linear regression. The SVM and MLFN with 31 nodes (MLFN-31) have the lowest RMSEs (0.11 and 0.10 respectively) and both having the prediction accuracy of 100% (under the tolerance of 30%). However, in our machines, the MLFN-31 requires 2 h for model training, while the SVM only needs about one second, which is also the shortest training time among the results in Table 1. Therefore, the SVM can be defined as the most suitable model for the prediction using the data provided by Fedele *et al.* [1].

The RMSEs shown in Table 2 and Figure 4b are comparatively higher than those in Table 1. Additionally, in Table 2, the RMSEs and training times of ANNs are comparatively higher than those of the SVM (RMSE: 6.09; training time: 0:00:01). The linear regression has the highest RMSE when testing (86.33). It can be apparently seen that the SVM is the most suitable model for the prediction using the data provided by Ihmels *et al.* [24].

In Table 3 and Figure 4c, the RMSE of the SVM is relatively higher than those of GRNN and MLFNs with low numbers of nodes. The MLFN with two nodes (MLFN-2) has the lowest RMSE (2.72) and a comparatively good prediction accuracy (100%, under the tolerance of 30%) among all models in Table 3 and, also, the training time of the MLFN-2 is comparatively short (0:01:13). Interestingly, when the numbers of nodes increase to 34 and 35, their corresponding prediction accuracies decrease to only 11.8%. This is because of the over-fitting phenomenon during the training of ANNs when the number of hidden nodes is relatively too high. Therefore, we can define that the MLFN-2 is the most suitable model for the prediction using the data provided by Klomfar *et al.* [25].

Table 4. Prediction models using experimental data by all the three experiment reports [1,24,25].

Model Type	RMSE (for Testing)	Training Time	Prediction Accuracy
Linear Regression	96.42	0:00:01	93.0%
SVM	15.79	0:00:02	99.2%
GRNN	92.33	0:00:02	93.0%
MLFN 2 Nodes	39.70	0:06:50	96.1%
MLFN 3 Nodes	25.03	0:08:36	97.7%
MLFN 4 Nodes	22.65	0:10:06	99.2%
MLFN 5 Nodes	73.84	0:13:49	93.0%
MLFN 6 Nodes	23.64	0:17:26	99.2%
MLFN 7 Nodes	65.74	0:14:39	93.8%
MLFN 8 Nodes	55.32	0:16:18	93.8%
. . .	. . .	. . .	. . .
MLFN 29 Nodes	164.54	0:52:29	89.1%
MLFN 30 Nodes	136.96	0:37:38	89.8%
MLFN 31 Nodes	168.13	0:41:35	89.1%
MLFN 32 Nodes	88.25	0:50:43	93.0%
MLFN 33 Nodes	143.65	2:30:12	89.8%
MLFN 34 Nodes	163.78	1:00:17	89.1%
MLFN 35 Nodes	166.92	0:44:16	89.1%

Table 4 and Figure 4d show that the SVM has the lowest RMSE (15.79), shortest training time (2 s), and highest prediction accuracy (99.2%). However, it is significant that the best predicted result presented in Table 4 and Figure 4d has a higher RMSE than those in Tables 1–3. A possible explanation of this phenomenon is that experimental details in different experiments may generate different deviations when acquiring the compressed liquid density of R227ea because the three groups of data come from three different research groups in different years [1,24,25]. Therefore, the combination of three groups of experimental data may generate additional noise, leading to deviations in training processes and, hence, the tested results have higher RMSEs. However, it should be noted that although the results of the best model here have higher RMSE than those in Tables 1–3 these testing results are still acceptable and it is also far more precise than the RMSEs generated by the theoretical equation of state.

3.2. Evaluation of Models

3.2.1. Comparison between Machine Learning Models and the Equation of State

To make comparisons among machine learning models and the theoretical model, we should first compare the RMSEs of different models (Table 5). Results show that the best machine learning models we have chosen in the four experimental groups are all apparently more precise than those results calculated by the Song and Mason equation, with lower RMSEs. The predicted values in the testing sets are generally highly close to their actual values in all the four machine learning models (Figure 5). It should be noted that experimental results provided by Fedele *et al.* [1] are generally more precise than the other two groups of experimental results [24,25], according to the generalized Tait equation [1,49]. Additionally, the testing RMSE of the SVM for the data provided by Fedele *et al.* [1] is the lowest during Table 5. One possible reason is that data provided by Fedele *et al.* [1] may have less experimental errors due to a well-developed measurement method, leading to better training effects, which indicates that data provided by Fedele *et al.* [1] is a good sample for training in practical predictions.

Table 5. RMSEs of different models.

Item	RMSE in Training	RMSE in Testing
SVM for data provided by Fedele *et al.* [1]	N/A	0.11
SVM for data provided by Ihmels *et al.* [24]	N/A	6.09
MLFN-2 for data provided by Klomfar *et al.* [25]	11.81	2.72
SVM for all data [1,24,25]	N/A	15.79
Theoretical calculation for data provided by Fedele *et al.* [1]	N/A	196.26
Theoretical calculation for data provided by Ihmels *et al.* [24]	N/A	372.54
Theoretical calculation for data provided by Klomfar *et al.* [25]	N/A	158.54

3.2.2. Comparison between Conventional Measurement Methods and Machine Learning

Advanced conventional approach for measuring the compressed liquid density of R227ea requires a series of apparatus connecting to be an entire system (Figure 6) [1]. However, the measurement requires time and a series of complex operations, which constraints its applicability. Additionally, the purchase and installation of the apparatus of conventional methods require too much manpower and resources, which indicates that it can only be used for acquiring extremely precise values. In contrast, machine learning models can make precise predictions based on the trained data set and give robust responses with a large number of trained data. Users can only input the new measured data of temperature and pressure and the precise predicted results can be automatically outputted by an appropriate machine learning model. Once the models are developed, new predicted data can be acquired in a very quick way, saving time and manpower. More importantly, it only needs a decent computer and no other apparatus is required anymore.

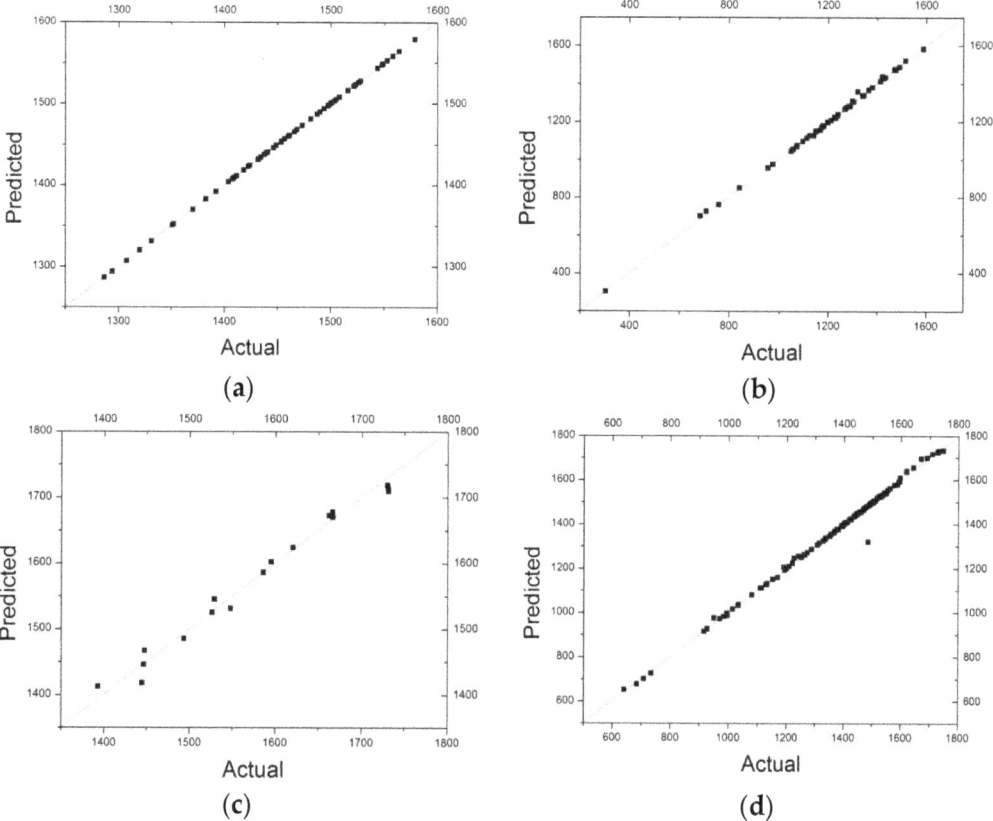

Figure 5. Predicted values *versus* actual values in testing processes using machine learning models. (**a**) The SVM for data provided by Fedele *et al.* [1]; (**b**) the SVM for data provided by Ihmels *et al.* [24]; (**c**) the MLFN-2 for data provided by Klomfar *et al.* [25]; and (**d**) the SVM for data provided by all the three experimental reports [1,24,25].

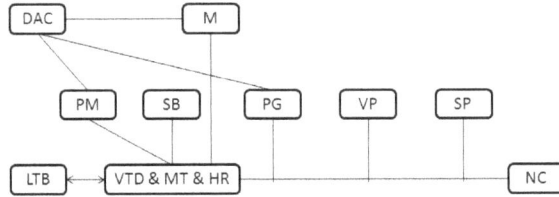

Figure 6. Apparatus scheme of density measuring for R227ea [1]. VTD represents the vibrating tube densimeter; PM represents the frequency meter; DAC represents the data acquisition and control; MT represents the temperature measurement sensor; M represents the multi-meter; LTB represents the liquid thermostatic bath; HR represents the heating resistance; SB represents the sample bottle; PG represents the pressure gauge; VP represents the vacuum pump; SP represents the syringe pump; NC represents the cylinder.

4. Conclusions

This study is a case study on the prediction of compressed liquid density of refrigerants, using R227ea as a typical example. To precisely acquire the densities of R227ea under different temperatures and pressures, existing measurements require complex apparatus and operations, wasting too much manpower and resources. Therefore, finding a method to predict the compressed liquid density directly is a good way to estimate the numerical values without tedious experiments. To provide a convenient methodology for predictions, a comparative study among different possible models is necessary [26,27,34,35]. Here, we used the Song and Mason equation, SVM, and ANNs to develop theoretical and machine learning models, respectively, for predicting the compressed liquid densities of R227ea. Results show that, compared to the Song and Mason equation, machine learning methods can better generate precise predicted results based on the experimental data. The SVMs are shown to be the best models for predicting the experimental results given by Fedele *et al.* [1], Ihmels *et al.* [24], and the combination of all the three experimental results [1,24,25]. The MLFN-2 is shown to be the best model for predicting the experimental results reported by Klomfar *et al.* [25]. It is also recommended that practical predictions can refer to the model developed with the training of experimental results reported by Fedele *et al.* [1] due to its more precise experimental results using advanced apparatus. Once a proper model is defined after model training and error analysis (such as the SVM for data provided by Fedele *et al.* in this case study), we can only input the easily-measured temperature and pressure, and then acquire the compressed liquid density of R227ea directly. Compared to experimental methods, machine learning can "put things right once and for all" with proper experimental data for model training. This study successfully shows that, in practical applications, users can only acquire the temperature and pressure of the measured R227ea and the density can be outputted by the developed appropriate model without additional operations. It should be noted that the target of this study is not to replace the traditional experimental works, but to give an alternative method for scientists and technicians to estimate the values as precise as possible in a limited time.

Author Contributions: Hao Li did the mathematical and modeling works. Xindong Tang simulated the theoretical density surface. Run Wang, Fan Lin, Zhijian Liu and Kewei Cheng joined the discussions and wrote the paper. This work is finished before Hao Li works at The University of Texas at Austin.

Conflicts of Interest: The authors declare no conflict of interest.

References

1.   Fedele, L.; Pernechele, F.; Bobbo, S.; Scattolini, M. Compressed liquid density measurements for 1,1,1,2,3,3,3-heptafluoropropane (R227ea). *J. Chem. Eng. Data* **2007**, *52*, 1955–1959.

2.   Garnett, T. Where are the best opportunities for reducing greenhouse gas emissions in the food system (including the food chain)? *Food Policy* **2011**, *36*, 23–23.

3.   Gholamalizadeh, E.; Kim, M.H. Three-dimensional CFD analysis for simulating the greenhouse effect in solar chimney power plants using a two-band radiation model. *Renew. Energy* **2014**, *63*, 498–506.

4.   Kang, S.M.; Polvani, L.M.; Fyfe, J.C.; Sigmond, M. Impact of polar ozone depletion on subtropical precipitation. *Science* **2011**, *332*, 951–954.

5.   Norval, M.; Lucas, R.M.; Cullen, A.P.; de Gruijl, F.R.; Longstreth, J.; Takizawa, Y.; van der Leun, J.C. The human health effects of ozone depletion and interactions with climate change. *Photochem. Photobiol. Sci.* **2011**, *10*, 199–225.

6.   Sun, J.; Reddy, A. Optimal control of building HVAC&R systems using complete simulation-based sequential quadratic programming (CSB-SQP). *Build. Environ.* **2005**, *40*, 657–669.

7.   T'Joen, C.; Park, Y.; Wang, Q.; Sommers, A.; Han, X.; Jacobi, A. A review on polymer heat exchangers for HVAC&R applications. *Int. J. Refrig.* **2009**, *32*, 763–779.

8.   Ladeinde, F.; Nearon, M.D. CFD applications in the HVAC and R industry. *Ashrae J.* **1997**, *39*, 44–48.

9.   Yang, Z.; Tian, T.; Wu, X.; Zhai, R.; Feng, B. Miscibility measurement and evaluation for the binary refrigerant mixture isobutane (R600a) + 1,1,1,2,3,3,3-heptafluoropropane (R227ea) with a mineral oil. *J. Chem. Eng. Data* **2015**, *60*, 1781–1786.

10.   Coquelet, C.; Richon, D.; Hong, D.N.; Chareton, A.; Baba-Ahmed, A. Vapour-liquid equilibrium data for the difluoromethane + 1,1,1,2,3,3,3-heptafluoropropane system at temperatures from 283.20 to 343.38 K and pressures up to 4.5 MPa. *Int. J. Refrig.* **2003**, *26*, 559–565.

11.   Fröba, A.P.; Botero, C.; Leipertz, A. Thermal diffusivity, sound speed, viscosity, and surface tension of R227ea (1,1,1,2,3,3,3-heptafluoropropane). *Int. J. Thermophys.* **2006**, *27*, 1609–1625.

12. Angelino, G.; Invernizzi, C. Experimental investigation on the thermal stability of some new zero ODP refrigerants. *Int. J. Refrig.* **2003**, *26*, 51–58.

13. Kruecke, W.; Zipfel, L. Foamed Plastic Blowing Agent; Nonflammable, Low Temperature Insulation. U.S. Patent No. 6,080,799, 27 June 2000.

14. Carlos, V.; Berthold, S.; Johann, F. Molecular dynamics studies for the new refrigerant R152a with simple model potentials. *Mol. Phys. Int. J. Interface Chem. Phys.* **1989**, *68*, 1079–1093.

15. Fermeglia, M.; Pricl, S. A novel approach to thermophysical properties prediction for chloro-fluoro-hydrocarbons. *Fluid Phase Equilibria* **1999**, *166*, 21–37.

16. Lísal, M.; Budinský, R.; Vacek, V.; Aim, K. Vapor-Liquid equilibria of alternative refrigerants by molecular dynamics simulations. *Int. J. Thermophys.* **1999**, *20*, 163–174.

17. Song, Y.; Mason, E.A. Equation of state for a fluid of hard convex bodies in any number of dimensions. *Phys. Rev. A* **1990**, *41*, 3121–3124.

18. Barker, J.A.; Henderson, D. Perturbation theory and equation of state for fluids. II. A successful theory of liquids. *J. Chem. Phys.* **1967**, *47*, 4714–4721.

19. Weeks, J.D.; Chandler, D.; Andersen, H.C. Role of repulsive forces in determining the equilibrium structure of simple liquids. *J. Chem. Phys.* **1971**, *54*, 5237–5247.

20. Mozaffari, F. Song and mason equation of state for refrigerants. *J. Mex. Chem. Soc.* **2014**, *58*, 235–238.

21. Bottou, L. From machine learning to machine reasoning. *Mach. Learn.* **2014**, *94*, 133–149.

22. Domingos, P.A. Few useful things to know about machine learning. *Commun. ACM* **2012**, *55*, 78–87.

23. Alpaydin, E. *Introduction to Machine Learning*; MIT Press: Cambridge, MA, USA, 2014.

24. Ihmels, E.C.; Horstmann, S.; Fischer, K.; Scalabrin, G.; Gmehling, J. Compressed liquid and supercritical densities of 1,1,1,2,3,3,3-heptafluoropropane (R227ea). *Int. J. Thermophys.* **2002**, *23*, 1571–1585.

25. Klomfar, J.; Hruby, J.; Sÿifner, O. measurements of the (T,p,ρ) behaviour of 1,1,1,2,3,3,3-heptafluoropropane (refrigerant R227ea) in the liquid phase. *J. Chem. Thermodyn.* **1994**, *26*, 965–970.

26. De Giorgi, M.G.; Campilongo, S.; Ficarella, A.; Congedo, P.M. Comparison between wind power prediction models based on wavelet decomposition with Least-Squares Support Vector Machine (LS-SVM) and Artificial Neural Network (ANN). *Energies* **2014**, *7*, 5251–5272.

27. De Giorgi, M.G.; Congedo, P.M.; Malvoni, M.; Laforgia, D. Error analysis of hybrid photovoltaic power forecasting models: A case study of mediterranean climate. *Energy Convers. Manag.* **2015**, *100*, 117–130.

28. Moon, J.W.; Jung, S.K.; Lee, Y.O.; Choi, S. Prediction performance of an artificial neural network model for the amount of cooling energy consumption in hotel rooms. *Energies* **2015**, *8*, 8226–8243.

29. Liu, Z.; Liu, K.; Li, H.; Zhang, X.; Jin, G.; Cheng, K. Artificial neural networks-based software for measuring heat collection rate and heat loss coefficient of water-in-glass evacuated tube solar water heaters. *PLoS ONE* **2015**, *10*, e0143624.

30. Zhang, Y.; Yang, J.; Wang, K.; Wang, Z. Wind power prediction considering nonlinear atmospheric disturbances. *Energies* **2015**, *8*, 475–489.

31. Leila, M.A.; Javanmardi, M.; Boushehri, A. An analytical equation of state for some liquid refrigerants. *Fluid Phase Equilib.* **2005**, *236*, 237–240.

32. Zhong, X.; Li, J.; Dou, H.; Deng, S.; Wang, G.; Jiang, Y.; Wang, Y.; Zhou, Z.; Wang, L.; Yan, F. Fuzzy nonlinear proximal support vector machine for land extraction based on remote sensing image. *PLoS ONE* **2013**, *8*, e69434.

33. Rebentrost, P.; Mohseni, M.; Lloyd, S. Quantum support vector machine for big data classification. *Phys. Rev. Lett.* **2014**, *113*.

34. Li, H.; Leng, W.; Zhou, Y.; Chen, F.; Xiu, Z.; Yang, D. Evaluation models for soil nutrient based on support vector machine and artificial neural networks. *Sci. World J.* **2014**, *2014*.

35. Liu, Z.; Li, H.; Zhang, X.; Jin, G.; Cheng, K. Novel method for measuring the heat collection rate and heat loss coefficient of water-in-glass evacuated tube solar water heaters based on artificial neural networks and support vector machine. *Energies* **2015**, *8*, 8814–8834.

36. Kim, D.W.; Lee, K.Y.; Lee, D.; Lee, K.H. A kernel-based subtractive clustering method. *Pattern Recognit. Lett.* **2005**, *26*, 879–891.

37. Fan, R.E.; Chang, K.W.; Hsieh, C.J.; Wang, X.R.; Lin, C.J. Liblinear: A library for large linear classification. *J. Mach. Learn. Res.* **2008**, *9*, 1871–1874.

38. Guo, Q.; Liu, Y. ModEco: An integrated software package for ecological niche modeling. *Ecography* **2010**, *33*, 637–642.

39. Hopfield, J.J. Artificial neural networks. *IEEE Circuits Devices Mag.* **1988**, *4*, 3–10.

40. Yegnanarayana, B. *Artificial Neural Networks*; PHI Learning: New Delhi, India, 2009.

41. Dayhoff, J.E.; DeLeo, J.M. Artificial neural networks. *Cancer* **2001**, *91*, 1615–1635.

42. Chang, C.C.; Lin, C.J. LIBSVM: A library for support vector machines. *ACM Trans. Intell. Syst. Technol.* **2011**, *2*, 389–396.

43. Specht, D.F. A general regression neural network. *IEEE Trans. Neural Netw.* **1991**, *2*, 568–576.

44. Tomandl, D.; Schober, A. A Modified General Regression Neural Network (MGRNN) with new, efficient training algorithms as a robust "black box"-tool for data analysis. *Neural Netw.* **2001**, *14*, 1023–1034.

45. Specht, D.F. The general regression neural network—Rediscovered. *IEEE Trans. Neural Netw. Learn. Syst.* **1993**, *6*, 1033–1034.

46. Svozil, D.; Kvasnicka, V.; Pospichal, J. Introduction to multi-layer feed-forward neural networks. *Chemom. Intell. Lab. Syst.* **1997**, *39*, 43–62.

47. Smits, J.; Melssen, W.; Buydens, L.; Kateman, G. Using artificial neural networks for solving chemical problems: Part I. Multi-layer feed-forward networks. *Chemom. Intell. Lab. Syst.* **1994**, *22*, 165–189.

48. Ilonen, J.; Kamarainen, J.K.; Lampinen, J. Differential evolution training algorithm for feed-forward neural networks. *Neural Process. Lett.* **2003**, *17*, 93–105.

49. Thomson, G.H.; Brobst, K.R.; Hankinson, R.W. An improved correlation for densities of compressed liquids and liquid mixtures. *AICHE J.* **1982**, *28*, 671–676.

Determination of Optimal Initial Weights of an Artificial Neural Network by Using the Harmony Search Algorithm: Application to Breakwater Armor Stones

Anzy Lee, Zong Woo Geem and Kyung-Duck Suh

Abstract: In this study, an artificial neural network (ANN) model is developed to predict the stability number of breakwater armor stones based on the experimental data reported by Van der Meer in 1988. The harmony search (HS) algorithm is used to determine the near-global optimal initial weights in the training of the model. The stratified sampling is used to sample the training data. A total of 25 HS-ANN hybrid models are tested with different combinations of HS algorithm parameters. The HS-ANN models are compared with the conventional ANN model, which uses a Monte Carlo simulation to determine the initial weights. Each model is run 50 times and the statistical analyses are conducted for the model results. The present models using stratified sampling are shown to be more accurate than those of previous studies. The statistical analyses for the model results show that the HS-ANN model with proper values of HS algorithm parameters can give much better and more stable prediction than the conventional ANN model.

Reprinted from *Appl. Sci.* Cite as: Lee, A.; Geem, Z.W.; Suh, K.-D. Determination of Optimal Initial Weights of an Artificial Neural Network by Using the Harmony Search Algorithm: Application to Breakwater Armor Stones. *Appl. Sci.* **2016**, *6*, 164.

1. Introduction

Artificial neural network (ANN) models have been widely used for prediction and forecast in various areas including finance, medicine, power generation, water resources and environmental sciences. Although the basic concept of artificial neurons was first proposed in 1943 [1], applications of ANNs have blossomed after the introduction of the back-propagation (BP) training algorithm for feedforward ANNs in 1986 [2], and the explosion in the capabilities of computers accelerated the employment of ANNs. The ANN models have also been used in various coastal and nearshore research [3–10].

An ANN model is a data-driven model aiming to mimic the systematic relationship between input and output data by training the network based on a large amount of data [11]. It is composed of the information-processing units called neurons, which are fully connected with different weights indicating the strength of the relationships between input and output data. Biases are also necessary to increase or decrease the net input of the neurons [12]. With the randomly selected initial weights and biases, the neural network cannot accurately estimate the required output. Therefore, the weights and biases are continuously modified by the so-called training so that the difference between the model output and target (observed) value becomes small. To train the network, the error function is defined as the sum of the squares of the differences. To minimize the error function, the BP training approach generally uses a gradient descent algorithm [11]. However, it can give a local minimum value of the error function as shown in Figure 1, and it is sensitive to the initial weights and biases. In other words, the gradient descent method is prone to giving a local minimum or maximum value [13,14]. If the initial weights and biases are fortunately selected to be close to the values that give the global minimum of the error function, the global minimum would be found by the gradient method. On the other hand, as expected in most cases, if they are selected to be far from the optimal values as shown by 'Start' in Figure 1, the final destination would be the local minimum that is marked by 'End' in the figure. As a consequence of local minimization, most ANNs provide erroneous results.

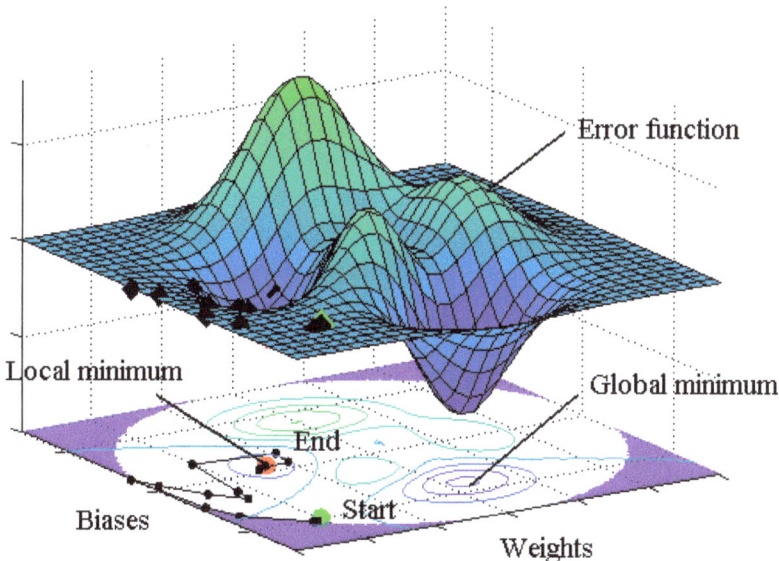

Figure 1. Illustration of local minimization problem.

To find the optimal initial weights and biases that lead into the global minimum of the error function, a Monte-Carlo simulation is often used, which, however, takes a long computation time. Moreover, even if we find the global optimal weights and biases by the simulation, they cannot be reproduced by the general users of the ANN model. Research has been performed to reveal and overcome the problem of local minimization in the ANN model. Kolen and Pollack [15] demonstrated that the BP training algorithm has large dependency on the initial weights by performing a Monte Carlo simulation. Yam and Chow [16] proposed an algorithm based on least-squares methods to determine the optimal initial weights, showing that the algorithm can reduce the model's dependency on the initial weights. Recently, genetic algorithms have been applied to find the optimal initial weights of ANNs and to improve the model accuracy [17–19]. Ensemble methods have also been implemented to enhance the accuracy of the model. They are also shown to overcome the dependency of the ANN model not only on the initial weights but also on training algorithms and data structure [20–23].

In this study, we employ the harmony search (HS) algorithm to find the near-global optimal initial weights of ANNs. It is a music-based metaheuristic optimization algorithm developed by Geem *et al.* [24] and has been applied to many different optimization problems such as function optimization, design of water distribution networks, engineering optimization, groundwater modeling, model parameter calibration, *etc.* The structure of the HS algorithm is much simpler than other metaheuristic algorithms. In addition, the intensification procedure conducted by the HS algorithm encourages speeding up the convergence by exploiting the history and experience in the search process. Thus, the HS algorithm in this study is expected to efficiently find the near-global optimal initial weights of the ANN. We develop an HS-ANN hybrid model to predict the stability number of armor stones of a rubble mound structure, for which a great amount of experimental data is available and thus several pieces of research using ANN models have been performed previously. The developed HS-ANN model is compared with the conventional ANN model without the HS algorithm in terms of the capability to find the global minimum of an error function. In the following section, previous studies for estimation of stability number are described; then, the HS-ANN model is developed in Section 3; the performance of the developed model is described in Section 4; and, finally, major conclusions are drawn in Section 5.

2. Previous Studies for Estimation of Stability Number

A breakwater is a port structure that is constructed to provide a calm basin for ships and to protect port facilities from rough seas. It is also used to protect the port area from intrusion of littoral drift. There are two basic types of breakwater: rubble mound breakwater and vertical breakwater. The cross section of a typical rubble mound breakwater is illustrated in Figure 2. To protect the rubble mound structure from severe erosion due to wave attack, an armor layer is placed on the seaward side of the structure. The stability of the armor units is measured by a dimensionless number, so-called stability number, which is defined as

$$N_s \equiv \frac{H_s}{\Delta D_{n50}},\tag{1}$$

where H_s is the significant wave height in front of the structure, $\Delta = \rho_s/\rho_w - 1$ is the relative mass density, ρ_s and ρ_w are the mass densities of armor unit and water, respectively, and D_{n50} is the nominal size of the armor unit. As shown in Equation (1), the stability number is defined as the ratio of the significant wave height to the size of armor units. A larger stability number, therefore, signifies that the armor unit with that size is stable against higher waves, that is, the larger the stability number, the more stable the armor units against waves.

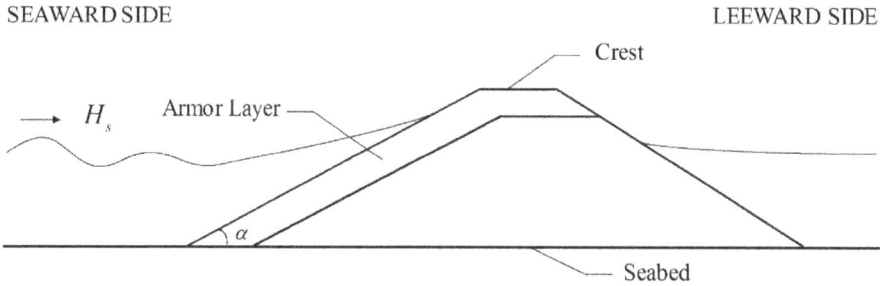

Figure 2. Typical cross-section of rubble mound breakwater.

To estimate the stability number, it is required to determine the relationship between the stability number and other variables which would describe the characteristics of waves and structure. Hudson [25] proposed an empirical formula:

$$N_s = (K_D \cot \alpha)^{1/3},\tag{2}$$

where α is the angle of structure slope measured from horizontal, and K_D is the stability coefficient which depends on the shape of the armor unit, the location at the structure (*i.e.*, trunk or head), placement method, and whether the structure is subject to breaking wave or non-breaking wave. The Hudson formula is simple, but it has been found to have a lot of shortcomings.

To solve the problems of the Hudson formula, Van der Meer [26] conducted an extensive series of experiments including the parameters which have significant effects on armor stability. Based on the experimental data, empirical formulas were proposed by Van der Meer [26,27] as follows:

$$N_s = \frac{1}{\sqrt{\xi_m}} \left[6.2 P^{0.18} \left(\frac{S}{\sqrt{N_w}} \right)^{0.2} \right] \text{ for } \xi_m < \xi_c, \qquad (3a)$$

$$N_s = 1.0 P^{-0.13} \left(\frac{S}{\sqrt{N_w}} \right)^{0.2} \sqrt{\cot \alpha} \, \xi_m^P \text{ for } \xi_m \geq \xi_c, \qquad (3b)$$

where $\xi_m = \tan \alpha / \sqrt{2\pi H_s / g T_m{}^2}$ is the surf similarity parameter based on the average wave period, T_m, $\xi_c = \left(6.2 P^{0.31} \sqrt{\tan \alpha} \right)^{1/(P+0.5)}$ is the critical surf similarity parameter indicating the transition from plunging waves to surging waves, P is the permeability of the core of the structure, N_w is the number of waves during a storm, and $S = A/D_{n50}^2$ (where A is the eroded cross-sectional area of the armor layer) is the damage level which is given depending on the degree of damage, e.g., onset of damage or failure.

On the other hand, with the developments in machine learning, various data-driven models have been developed based on the experimental data of Van der Meer [26]. A brief summary is given here only for the ANN models. Mase et al. [3] constructed an ANN by using randomly selected 100 experimental data of Van der Meer [26] for training the network. The total number of experimental data excluding the data of low-crested structures was 579. In the test of the ANN, they additionally used the 30 data of Smith et al. [28]. They employed six input variables: P, N_w, S, ξ_m, h/H_s, and the spectral shape parameter SS, where h is the water depth in front of the structure. Kim and Park [6] followed the approach of Mase et al. [3] except that they used 641 data including low-crested structures. Since, in general, the predictive ability of an ANN is improved as the dimension of input variables increases, they split the surf similarity parameter into structure slope and wave steepness, and the wave steepness further into wave height and period. Note that the surf similarity parameter ξ_m consists of structure slope, wave height and period as shown in its definition below Equation (3), where $H_s/L_m = H_s/(g T_m^2/2\pi)$ is the wave steepness. They showed that the ANN gives better performance as the input dimension is increased. On the other hand, Balas et al. [9] used principal component analysis (PCA) based on 554 data of Van der Meer [26] to develop hybrid ANN models. They created four different models by reducing the data from 544 to 166 by using PCA or by using the principal components as the input variables of the ANN. Table 1 shows the correlation coefficients of previous studies, which will be compared with those of the present study later.

Table 1. Correlation coefficients of different empirical formula or ANN models.

Author	Correlation Coefficient	Number of Data	Remarks
Van der Meer [27]	0.92 (Mase *et al.* [3])	579	Empirical formula, Equation (3) in this paper
Mase *et al.* [3]	0.91	609	Including data of Smith *et al.* [28]
Kim and Park [6]	0.902 to 0.952	641	Including data of low-crested structures
Balas *et al.* [9]	0.906 to 0.936	554	ANN-PCA hybrid models

3. Development of an HS-ANN Hybrid Model

3.1. Sampling of Training Data of ANN Model

The data used for developing an ANN model is divided into two parts: the training data for training the model and the test data for verifying or testing the performance of the trained model. The training data should be sampled to represent the characteristics of the population. Otherwise, the model would not perform well for the cases that had not been encountered during the training. For example, if a variable of the training data consists of only relatively small values, the model would not perform well for large values of the variable because the model has not experienced the large values and vice versa. Therefore, in general, the size of the training data is taken to be larger than that of the test data. In the previous studies of Mase *et al.* [3] and Kim and Park [6], however, only 100 randomly sampled data were used for training the models, which is much smaller than the total number of data, 579 or 641. This might be one of the reasons why the ANN models do not show superior performance compared with the empirical formula (see Table 1).

To overcome this problem, the stratified sampling method is used in this study to sample 100 training data as in the previous studies while using the remaining 479 data to test the model. The key idea of stratified sampling is to divide the whole range of a variable into many sub-ranges and to sample the data so that the probability mass in each sub-range becomes similar between sample and population. Since a number of variables are involved in this study, the sampling was done manually. There are two kinds of statistical tests to evaluate the performance of stratified sampling, *i.e.*, parametric and non-parametric tests. Since the probability mass function of each variable in this study does not follow the normal distribution, the chi-square (χ^2) test is used, which is one of the non-parametric tests. The test is fundamentally based on the error between the assumed and observed probability densities [29]. In the test, each of the range of the n observed data is divided into m sub-ranges. In addition, the number of frequencies (n_i) of the variable in the ith sub-range is counted. Furthermore, the observed frequencies (n_i, $i = 1$ to m) and

23

the corresponding theoretical frequencies (e_i, $i = 1$ to m) of an assumed distribution are compared. As the total sample points n tends to ∞, it can be shown [30] that the quantity, $\sum_{i=1}^{m} (n_i - e_i)^2/e_i$, approaches the χ^2 distribution with $f = m - 1 - k$ degree of freedom, where k is the number of parameters in the assumed distribution. k is set to zero for non-normal distribution. The observed distribution is considered to follow the assumed distribution with the level of significance σ if

$$\sum_{i=1}^{m} \frac{(n_i - e_i)^2}{e_i} < c_{1-\sigma,f}, \tag{4}$$

where $c_{1-\sigma,f}$ indicates the value of the χ^2 distribution with f degree of freedom at the cumulative mass of $1 - \sigma$. In this study, a 5% level of significance is used.

Table 2 shows the input and output variables in the ANN model. The surf similarity parameter was split into $\cot\alpha$, H_s, and T_p as done by Kim and Park [6]. The peak period, T_p, was used instead of T_m because it contains the information about spectral peak as well as mean wave period. The neural network can deal with qualitative data by assigning the values to them. The permeability coefficients of impermeable core, permeable core, and homogeneous structure are assigned to 0.1, 0.5, and 0.6, respectively, as done by Van der Meer [27]. On the other hand, the spectral shapes of narrowband, medium-band (*i.e.*, Pierson-Moskowitz spectrum), and wideband are assigned to 1.0, 0.5, and 0, as done by Mase *et al.* [3]. To perform the chi-square test, the range of each variable was divided into eight to 11 sub-ranges. The details of the test can be found in the thesis of Lee [31]. Here, only the residual chart calculated based on Equation (4) is presented in Table 3. Some variables are well distributed over the whole range, whereas some varies among a few sub-ranges (e.g., $P = 0.1$, 0.5, or 0.6). Table 3 shows that Equation (4) is satisfied for all the variables, indicating that the probability mass function of each variable of the training data is significant at a 5% level of significance. As an example, the probability mass functions of the training data and population of the damage level S are compared in Figure 3, showing that they are in good agreement.

Table 2. Input and output variables.

Input Variables	Output Variable
P, N_w, S, $\cot\alpha$, H_s, T_p, h/H_s, SS	N_s

Table 3. Residual chart of chi-square tests.

Range	N_s	P	N_w	S	$\cot\alpha$	H_s	T_p	h/H_s	SS
1	0.09	0.26	0.03	0.42	0.13	0.06	-	-	0.31
2	0.00	-	-	0.03	-	0.00	-	1.74	-
3	0.00	-	-	0.01	0.00	1.24	-	0.00	-
4	0.00	-	-	0.01	-	0.14	0.84	0.17	-
5	0.05	-	-	0.12	0.47	0.11	0.00	0.08	0.00
6	0.10	-	-	0.14	-	0.02	0.02	0.02	-
7	0.04	-	-	0.11	-	1.24	0.00	0.06	-
8	0.07	0.35	-	0.06	-	-	0.02	0.00	-
9	0.14	-	-	0.90	-	-	0.38	-	-
10	1.50	0.04	0.03	0.45	0.08	-	-	0.14	0.37
11	-	-	-	0.43	-	-	0.52	-	-
$\sum(n_i-e_i)^2/e_i$	1.99	0.64	0.06	2.67	0.68	2.81	1.77	2.20	0.69
f	9	2	1	10	3	6	6	7	2
$c_{1-\sigma,f}$	16.8	5.99	3.84	18.3	7.86	12.6	12.6	14.1	5.99

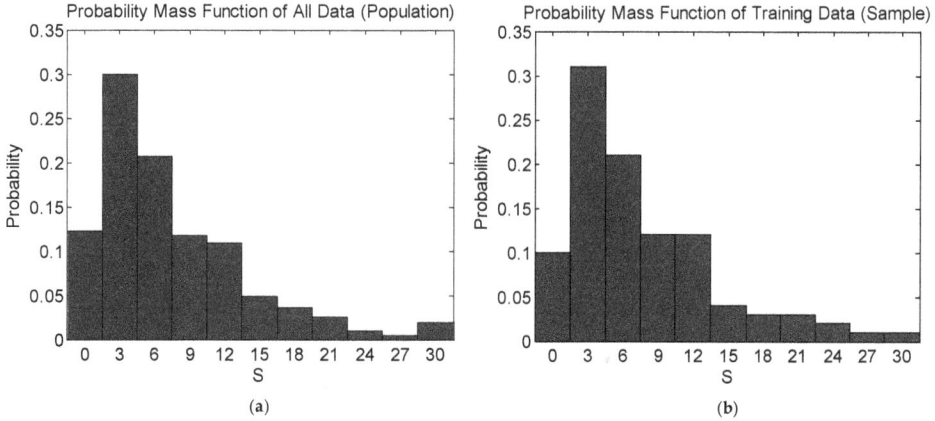

Figure 3. Probability mass functions of damage level S: (a) population; (b) training data.

3.2. ANN Model

The multi-perceptron is considered as an attractive alternative to an empirical formula in that it imitates the nonlinear relationship between input and output variables in a more simplified way. The model aims to obtain the optimized weights of the network using a training algorithm designed to minimize the error between the output and target variables by modifying the mutually connected weights. In this study, the multi-perceptron with one hidden layer is used as shown in Figure 4, where i is the number of input variables.

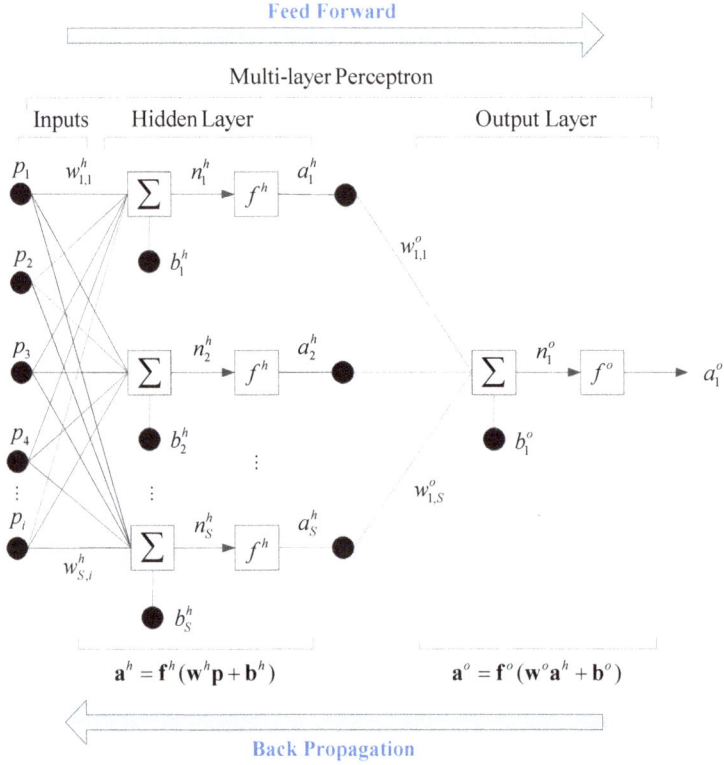

Figure 4. Network topography of ANN.

Firstly, for each of the input and output variables, the data are normalized so that all of the data are distributed in the range of [min, max] = [−1, 1]. This can be done by subtracting the average from the data values and rescaling the resulting values in such a way that the smallest and largest values become −1 and 1, respectively. Secondly, the initial weights in the hidden layer are set to have random values between −1 and 1, and the initial biases are all set to zero. The next step is to multiply the weight matrix by the input data, **p**, and add the bias so that

$$n_k^h = \sum_{j=1}^{J} w_{kj}^h p_j + b_k^h, \quad k = 1 \text{ to } K, \tag{5}$$

where J and K are the number of input variables and hidden neurons, respectively, and **p**, $\mathbf{b}^h$ and $\mathbf{w}^h$ are the input variable, bias, and weight in the hidden layer, respectively. The subscripts of the weight w_{kj}^h are written in such a manner that the first subscript denotes the neuron in question and the second one indicates the input variable to which the weight refers. The n_k^h calculated by Equation (5) is fed into an activation function, f^h, to calculate a_k^h. Hyperbolic tangent sigmoid function is used as the activation function so that

$$a_k^h = \frac{e^{n_k^h} - e^{-n_k^h}}{e^{n_k^h} + e^{-n_k^h}}. \tag{6}$$

In the output layer, the same procedure as that in the hidden layer is used except that only one neuron is used so that

$$n_1^o = \sum_{j=1}^{K} w_{1j}^o a_j^h + b_1^o, \tag{7}$$

and the linear activation function is used to calculate a_1^o so that

$$a_1^o = n_1^o. \tag{8}$$

The neural network with the randomly assigned initial weights and biases cannot accurately estimate the required output. Therefore, the weights and biases are modified by the training to minimize the difference between the model output and target (observed) values. To train the network, the error function is defined as

$$\varepsilon = ||\boldsymbol{\tau} - \mathbf{a}_1^o||^2, \tag{9}$$

where $||\ \ ||$ denotes a norm, and $\boldsymbol{\tau}$ is the target value vector to be sought. To minimize the error function, the Levenberg-Marquardt algorithm is used, which is the standard algorithm of nonlinear least-squares problems. Like other numeric minimization algorithms, the Levenberg-Marquardt algorithm is an iterative procedure. It necessitates a damping parameter μ, and a factor θ that is greater than one. In this study, $\mu = 0.001$ and $\theta = 10$ were used. If the squared error increases, then the damping is increased by successive multiplication by θ until the error decreases with a new damping parameter of $\mu\theta^k$ for some k. If the error decreases, the damping parameter is divided by θ in the next step. The training was stopped when the epoch reached 5000 or the damping parameter was too large for more training to be performed.

3.3. HS-ANN Hybrid Model

To find the initial weights of the ANN model that lead into the global minimum of the error function, in general, a Monte Carlo simulation is performed, that is, the training is repeated many times with different initial weights. The Monte Carlo simulation, however, takes a great computational time. In this section, we develop an HS-ANN model in which the near-global optimal initial weights are found by the HS algorithm.

The HS algorithm consists of five steps as follows [32].

Step 1. Initialization of the algorithm parameters

Generally, the problem of global optimization can be written as

$$\text{Minimize} \quad f(\mathbf{x})$$
$$\text{subject to} \quad x_i \in \mathbf{X}_i, \ i = 1, 2, ..., N, \tag{10}$$

where $f(\mathbf{x})$ is an objective function, $\mathbf{x}$ is the set of decision variables, and $\mathbf{X}_i$ is the set of possible ranges of the values of each decision variable, which can be denoted as $\mathbf{X}_i = \{x_i(1), x_i(2), ..., x_i(K)\}$ for discrete decision variables satisfying $x_i(1) < x_i(2) < \cdots < x_i(K)$ or for continuous decision variables. In addition, N is the number of decision variables and K is the number of possible values for the discrete variables. In addition, HS algorithm parameters exist that are required to solve the optimization problems: harmony memory size (HMS, number of solution vectors), harmony memory considering rate (HMCR), pitch adjusting rate (PAR) and termination criterion (maximum number of improvisation). HMCR and PAR are the parameters used to improve the solution vector.

Step 2. Initialization of harmony memory

The harmony memory (HM) matrix is composed of as many randomly generated solution vectors as the size of the HM, as shown in Equation (11). They are stored with the values of the objective function, $f(\mathbf{x})$, ascendingly:

$$\text{HM} = \begin{bmatrix} \mathbf{x}^1 \\ \mathbf{x}^2 \\ \vdots \\ \mathbf{x}^{\text{HMS}} \end{bmatrix}. \tag{11}$$

Step 3. Improvise a new harmony from the HM

A new harmony vector, $\mathbf{x}' = (x_1', x_2', ..., x_N')$, is created from the HM based on assigned HMCR, PAR, and randomization. For example, the value of the first decision variable (x_1') for the new vector can be selected from any value in the designated HM range, $x_1^1 \sim x_1^{\text{HMS}}$. In the same way, the values of other decision variables can be selected. The HMCR parameter, which varies between 0 and 1, is a possibility that the new value is selected from the HM as follows:

$$x'_i \leftarrow \begin{cases} x'_i \in \{x^1_i, x^2_i, ..., x^{HMS}_i\} & \text{with probability of HMCR} \\ x'_i \in \mathbf{X}_i & \text{with probability of } (1 - HMCR), \end{cases} \quad (12)$$

The HMCR is the probability of selecting one value from the historic values stored in the HM, and the value (1-HMCR) is the probability of randomly taking one value from the possible range of values. This procedure is analogous to the mutation operator in genetic algorithms. For instance, if a HMCR is 0.95, the HS algorithm would pick the decision variable value from the HM including historically stored values with a 95% of probability. Otherwise, with a 5% of probability, it takes the value from the entire possible range. A low memory considering rate selects only a few of the best harmonies, and it may converge too slowly. If this rate is near 1, most of the pitches in the harmony memory are used, and other ones are not exploited well, which does not lead to good solutions. Therefore, typically HMCR = 0.7 − 0.95 is recommended.

On the other hand, the HS algorithm would examine every component of the new harmony vector, $\mathbf{x}' = (x'_1, x'_2, ..., x'_N)$, to decide whether it has to be pitch-adjusted or not. In this procedure, the PAR parameter which sets the probability of adjustment for the pitch from the HM is used as follows:

$$\text{Pitch adjusting decision for } x'_i \leftarrow \begin{cases} \text{Yes} & \text{with probability of PAR} \\ \text{No} & \text{with probability of } (1 - PAR). \end{cases} \quad (13)$$

The pitch adjusting procedure is conducted only after a value is selected from the HM. The value (1−PAR) is the probability of doing nothing. To be specific, if the value of PAR is 0.1, the algorithm will take a neighboring value with 0.1 × HMCR probability. For example, if the decision for x'_i in the pitch adjustment process is Yes, and x'_i is considered to be $x_i(k)$, then the kth element in $\mathbf{X}_i$, or the pitch-adjusted value of $x_i(k)$, is changed into

$$\begin{aligned} x'_i &\leftarrow x_i(k + m) && \text{for discrete decision variables} \\ x'_i &\leftarrow x'_i + \alpha && \text{for continuous decision variables,} \end{aligned} \quad (14)$$

where m is the neighboring index, $m \in \{ ..., -2, -1, 1, 2, ...\}$, α is the value of $bw \times u(-1, 1)$, bw is an arbitrarily chosen distance bandwidth or fret width for the continuous variable, and $u(-1, 1)$ is a random number from uniform distribution with the range of $[-1, 1]$. If the pitch-adjusting rate is very low, because of the limitation in the exploration of a small subspace of the whole search space, it slows down the convergence of HS. On the contrary, if the rate is very high, it may cause the solution to scatter around some potential optima. Therefore, PAR $= 0.1 - 0.5$ is used in most applications. The parameters HMCR and PAR help the HS algorithm to search globally and locally, respectively, to improve the solution.

Step 4. Evaluate new harmony and update the HM

This step is to evaluate the new harmony and update the HM if necessary. Evaluating a new harmony means that the new harmony (or solution vector) is used in the objective function and the resulting functional value is compared with the solution vector in the existing HM. If the new harmony vector gives better performance than the worst harmony in the HM, evaluated in terms of the value of objective function, the new harmony would be included in the harmony memory and the existing worst harmony is eliminated from the harmony memory. In this study, the mean square error function is used as the objective function for both HS optimization and ANN training.

Step 5. Repeat Steps 3 and 4 until the termination criterion is satisfied

The iterations are terminated if the stop criterion is satisfied. If not, Steps 3 and 4 would be repeated. The pseudo-code of the HS algorithm is given in Figure 5. The initial weights optimized by the HS algorithm are then further trained by a gradient descent algorithm. This method is denoted as an HS-ANN model, and it can be expressed as the flow chart in Figure 6.

Harmony Search

begin
     Objective function f(x), $x = (x_1, x_2, \ldots, x_d)^T$
     Generate initial Harmony Memory (HM)
     Define harmony memory considering rate (HMCR)
     Define pitch adjusting rate PAR and other parameters
     while *(t<Max number of iterations)*
       while *(i<=number of variables)*
         if *(rand<HMCR), Choose a value from HM for the variable i*
           if *(rand<PAR), Adjust the pitch randomly within bounds*
         end if
         else *Generate new harmonies by randomization*
       end if
       end while
      Accept the new harmony (solution) if better
     end while
     Find the current best solution
 End

Figure 5. Pseudo-code of HS algorithm (modified from Geem [33]).

4. Result and Discussion

4.1. Assessment of Accuracy and Stability of the Models

In this section, the accuracy and stability are compared between the conventional ANN model without using the HS algorithm and the HS-ANN hybrid model. Both models were run 50 times, and the statistical analyses were conducted for the model results. Each of the HMCR and PAR of the HS algorithm were chosen to vary from 0.1 to 0.9 at intervals of 0.2, so a total of 25 HS-ANN models were tested. The models were used to estimate the stability number of rock armor for the experimental data of Van der Meer [26] for which the input and output variables are given as in Table 2. The 579 experimental data excluding the data of low-crested structures were used, as done by Mase *et al.* [3]. As described in Section 3.1, a hundred data sampled by the stratified sampling method were used to train the ANN, whereas the remaining 479 data were used to test the model.

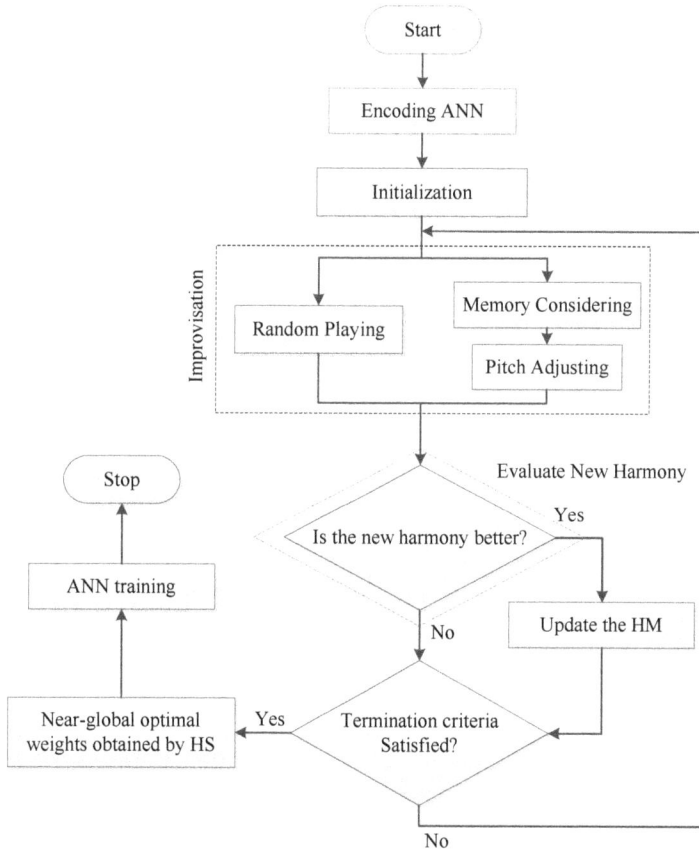

Figure 6. Flow chart of HS-ANN model.

The correlation coefficient (r) and index of agreement (I_a) between model output values and target values of the 479 test data are used to evaluate the performance of the models. First, to compare the accuracy of the developed models with those of previous studies (see Table 1), the maximum value of correlation coefficient among 50 runs of each model is presented in Table 4. For the results of the HS-ANN models, the rank is indicated by a superscript, and the largest two values are shaded. The largest correlation coefficient of the HS-ANN model is only slightly larger than that of the ANN model, but both of them are much greater than those of previous studies (see Table 1), probably because the stratified sampling was used in the present study.

Table 4. Maximum values of correlation coefficient.

	HMCR\PAR	0.1	0.3	0.5	0.7	0.9
	0.1	0.957	0.971	0.961	0.973^1	0.964
HS-ANN	0.3	0.959	0.967	0.970	0.972^3	0.960
	0.5	0.961	0.954	0.961	0.957	0.968
	0.7	0.968	0.973^2	0.959	0.967	0.970
	0.9	0.971^5	0.970	0.972^4	0.970	0.960
ANN				0.971		

Even though we used the correlation coefficient to compare the accuracy of our model results with those of previous studies, it is not a good indicator for model accuracy because it merely evaluates the linearity between observation and prediction but not the agreement between them. Hereinafter, we use the index of agreement introduced by Willmott [34] as a measure of the degree to which a model's predictions are error-free but not a measure of correlation between the observed and predicted variates. The index of agreement is given as

$$I_a = 1 - \frac{\sum_{i=1}^{N} (p_i - o_i)^2}{\sum_{i=1}^{N} [|p_i - \bar{o}| + |o_i - \bar{o}|]^2}, \tag{15}$$

where p_i and o_i denote the predicted and observed variates, and $\bar{o}$ is the mean of the observed variates. The values for I_a vary between 0 and 1.0, where 1.0 indicates perfect agreement between observations and predictions, and 0 connotes complete disagreement.

The statistical parameters used to measure the predictive ability and stability of the models are the average, standard deviation, and the minimum value of I_a. The larger the average, the better the overall predictive ability of the model. The smaller the standard deviation, the higher the stability of the model, that is, the less variability among the model outputs from different runs of the model. Lastly, the larger the minimum value of I_a, the larger the lower threshold of the predictive ability of the model. In summary, a large average and large minimum value of I_a

signify that the predictive ability of the model is excellent. On the other hand, a small standard deviation signifies that the model is stable.

The statistical parameters for the index of agreement are presented in Table 5. Again, for the results of the HS-ANN models, the rank is indicated by a superscript, and the largest (or smallest) two values are shaded. Even though the maximum values of I_a are also given, they are not discussed further because their variation is not so large to explain the difference of predictive ability or stability depending on the models. It is shown that the HS-ANN model gives the most excellent predictive ability and stability with HMCR = 0.7 and PAR = 0.5 or HMCR = 0.9 and PAR = 0.1. This result corresponds to Geem [33] who suggested that the optimal ranges of HMCR = 0.7–0.95 and PAR = 0.1–0.5. Comparing the statistical parameters between the best HS-ANN model and the ANN model, the HS-ANN model with proper values of HMCR and PAR can give much better and stable prediction than the conventional ANN model. In particular, the small value of standard deviation of the HS-ANN model indicates that the model is excellent in finding the global minimum of the error function.

4.2. Aspect of Transition of Weights

There are two major components in metaheuristic algorithms: diversification and intensification [33]. These two components seem to be contradicting each other, but balancing their combination is crucial and important to the success of a metaheuristic algorithm. In the HS algorithm, diversification is controlled by the pitch adjustment and randomization, whereas intensification is represented by the harmony memory considering rate. Therefore, in this section, the results of training of neural networks for two different cases are compared and examined, *i.e.*, the best combination and worst combination of HMCR and PAR. The case of the HS optimization with HMCR of 0.7 and PAR of 0.5 is chosen to be the best case (Case 1) since it has the largest average and smallest standard deviation of I_a. The worst case (Case 2) is the case of HMCR of 0.1 and PAR of 0.5, which has the smallest average and largest standard deviation of I_a. The optimization process of the HS algorithm regarding the weights of neural network is illustrated in Figures 7 and 8 for each case of parameter combination. Note that the results shown in these figures are those from one of the fifty runs described in the previous section. In the figures, each scatter plot indicates the relationship between calculated and observed stability numbers using (a) randomly selected initial weights; (b) optimal initial weights determined by HS algorithm; and (c) further trained weights after BP algorithm. The correlation coefficients and indices of agreement are also given.

Table 5. Statistical parameters for index of agreement.

	HMCR\PAR	0.1	0.3	0.5	0.7	0.9
				Average		
	HMCR\PAR	0.1	0.3	0.5	0.7	0.9
	0.1	0.885	0.926	0.872	0.843	0.905
HS-ANN	0.3	0.934	0.910	0.913	0.914	0.912
	0.5	0.913	0.929	0.929	0.934	0.929
	0.7	0.881	0.929	0.948^{1}	0.944^{3}	0.940^{4}
	0.9	0.948^{2}	0.935	0.913	0.937^{5}	0.892
ANN				0.804		

	HMCR\PAR	0.1	0.3	0.5	0.7	0.9
				Standard Deviation		
	HMCR\PAR	0.1	0.3	0.5	0.7	0.9
	0.1	0.205	0.120	0.245	0.277	0.158
HS-ANN	0.3	0.137	0.175	0.155	0.183	0.189
	0.5	0.178	0.101	0.104	0.073	0.138
	0.7	0.224	0.130	0.021^{1}	0.031^{3}	0.042^{5}
	0.9	0.023^{2}	0.110	0.168	0.031^{4}	0.200
ANN				0.317		

	HMCR\PAR	0.1	0.3	0.5	0.7	0.9
				Minimum Value		
	HMCR\PAR	0.1	0.3	0.5	0.7	0.9
	0.1	0.001	0.216	0.001	0.001	0.002
HS-ANN	0.3	0.002	0.001	0.029	0.004	0.002
	0.5	0.003	0.319	0.254	0.468	0.003
	0.7	0.001	0.051	0.889^{1}	0.852^{3}	0.710^{5}
	0.9	0.885^{2}	0.196	0.013	0.801^{4}	0.005
ANN				0.001		

	HMCR\PAR	0.1	0.3	0.5	0.7	0.9
				Maximum Value		
	HMCR\PAR	0.1	0.3	0.5	0.7	0.9
	0.1	0.978	0.985	0.980	0.987^{1}	0.981
HS-ANN	0.3	0.979	0.983	0.985	0.986	0.979
	0.5	0.980	0.977	0.980	0.978	0.984
	0.7	0.984	0.986^{2}	0.979	0.983	0.985
	0.9	0.985	0.985	0.986	0.985	0.980
ANN				0.985		

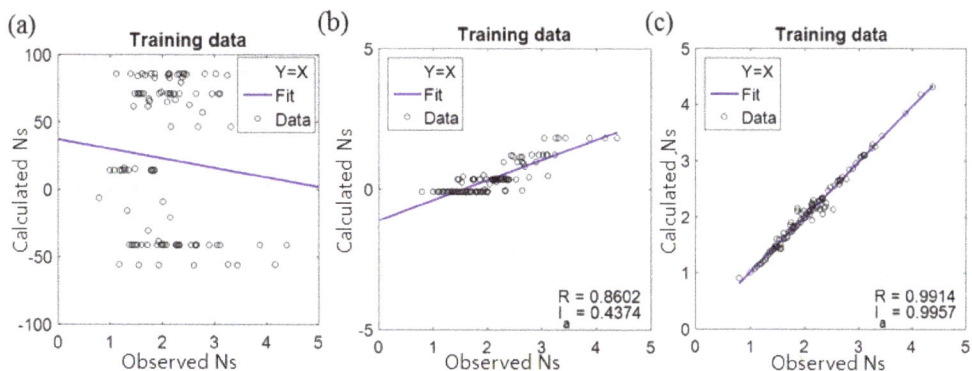

Figure 7. Scatter plots of calculated *versus* observed stability numbers for Case 1 (HMCR = 0.7; PAR = 0.5): (**a**) randomly selected initial weights; (**b**) optimal initial weights determined by HS algorithm; (**c**) further trained weights after BP algorithm.

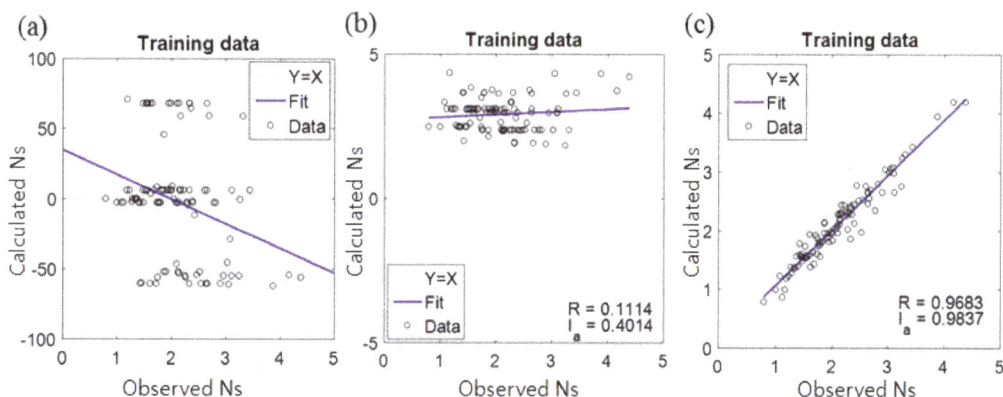

Figure 8. Same as Figure 7 but for Case 2 (HMCR = 0.1; PAR = 0.5).

The first graphs in Figures 7 and 8 show that the stability numbers calculated by using the randomly selected initial weights are distributed in a wide range from large negative values to large positive values and have very weak and negative correlation with the observed stability numbers. The second graphs show that the stability numbers calculated by using the optimal initial weights determined by the HS algorithm are distributed within the range of 0 to 5 as they are in the observation. Case 1 shows much better correlation between calculation and observation than the Case 2, whereas the index of agreement of Case 1 is only slightly better than that of Case 2. The calculated stability numbers in Case 1 show strong correlation with the observed ones, but they are underestimated as a whole. The third graphs after further training by the BP algorithm show very strong correlation and good agreement between calculation and observation. Even Case 2 (the worst case) shows quite high value of index of agreement compared with the values for HMCR = 0.1 and PAR = 0.5 in Table 5. Note that the results in Figures 7 and 8 are for training data, whereas those in Table 5 are for test data.

4.3. Computational Time

Most of the computational time of the conventional ANN model is used for the BP training of the model, whereas the HS-ANN model needs computational time for finding the optimal initial weights using the HS algorithm and then for finding the global minimum of the error function by the BP training. Lee [31] compared the computational time between the conventional ANN model and the HS-ANN models with various combinations of HMCR and PAR. Since the statistical characteristics of computational time do not show a big difference among different combinations of HMCR and PAR, here we only present the case of HMCR = 0.7 and PAR = 0.5 for which the HS-ANN model gives the most excellent predictive ability and stability. Table 6 shows the average and standard deviation (SD) of the computational times of the 50 runs of each model. The total computational time of the HS-ANN model is five to six times greater than that of the conventional ANN model. In spite of the greater computing time, it is worthwhile to use the HS-ANN model because it gives much more accurate and stable prediction than the conventional ANN model with a small number of simulations. It is interesting to note that the computing time for the BP training of the HS-ANN model is greater than that of the conventional ANN model probably because it takes a longer time to reach the global minimum which is smaller than the local minimums as shown in Figure 1. On the other hand, the standard deviation of the BP training of the HS-ANN model is smaller than that of the conventional ANN model because the HS-ANN model starts the BP training from the optimal initial weights whose variation is not so large. The standard deviation of the HS algorithm is very small because the maximum number of improvisation was set to 100,000.

Table 6. Statistical characteristics of computational time (unit = s).

Algorithm	HS-ANN Model		Conventional ANN Model	
	Average	SD	Average	SD
HS	285.6	7.8	-	-
BP	102.9	55.2	68.6	95.0
Total	385.5	55.7	68.6	95.0

5. Conclusion

In this study, an HS-ANN hybrid model was developed to predict the stability number of breakwater armor stones based on the experimental data of Van der Meer [26]. The HS algorithm was used to find the near-global optimal initial weights, which were then used in the BP training to find the true global minimum of the error function by the Levenberg-Marquardt algorithm. The stratified sampling was used to sample the training data. A total of 25 HS-ANN models were tested with five different values for both HMCR and PAR varying from 0.1 to 0.9 at intervals of 0.2. The HS-ANN models were compared with the conventional ANN model which uses a Monte Carlo simulation to determine the initial weights. The correlation coefficient and index of agreement were calculated to evaluate the performance of the models. Each model was run 50 times and the statistical analyses were conducted for the model results. The major findings are as follows:

1. The correlation coefficients of the present study were greater than those of previous studies probably because of the use of stratified sampling.
2. In terms of the index of agreement, the HS-ANN model gave the most excellent predictive ability and stability with HMCR = 0.7 and PAR = 0.5 or HMCR = 0.9 and PAR = 0.1, which correspond to Geem [33] who suggested the optimal ranges of HMCR = 0.7–0.95 and PAR = 0.1–0.5 for the HS algorithm.
3. The statistical analyses showed that the HS-ANN model with proper values of HMCR and PAR can give much better and more stable prediction than the conventional ANN model.
4. The HS algorithm was found to be excellent in finding the global minimum of an error function. Therefore, the HS-ANN hybrid model would solve the local minimization problem of the conventional ANN model using a Monte Carlo simulation, and thus could be used as a robust and reliable ANN model not only in coastal engineering but also other research areas.

In the future, the present HS-ANN model could be compared with other hybrid ANN models using different heuristic algorithms such as genetic algorithm (GA), particle swarm optimization (PSO), and Cuckoo Search (CS). Not only GA [18,35,36] but also PSO [37,38] and CS [39] have been applied for neural network training. Analyzing and comparing those hybrid ANN models would provide a way to find the most suitable heuristic algorithm for determining the optimal initial weights for ANN.

Acknowledgments: This research was supported by Basic Science Research Program through the National Research Foundation of Korea (NRF) funded by the Ministry of Science, ICT and Future Planning (NRF-2014R1A2A2A01007921). The Institute of Engineering Research and Entrepreneurship at Seoul National University provided research facilities for this work.

Author Contributions: A.L. conducted the research and wrote the paper; Z.W.G. advised on the use of HS algorithm; K.D.S. conceived, designed, and directed the research.

Conflicts of Interest: The authors declare no conflict of interest.

References

1. McCulloch, W.S.; Pitts, W. A logical calculus of the ideas immanent in nervous activity. *Bull. Math. Biol.* **1990**, *52*, 99–115.
2. Rumelhart, D.E.; Hinton, G.E.; Williams, R.J. Learning representations by back-propagating errors. *Nature* **1986**, *323*, 533–536.
3. Mase, H.; Sakamoto, M.; Sakai, T. Neural network for stability analysis of rubble-mound breakwaters. *J. Waterway Port Coast. Ocean Eng.* **1995**, *121*, 294–299.
4. Tsai, C.P.; Lee, T. Back-propagation neural network in tidal-level forecasting. *J. Waterway Port Coast. Ocean Eng.* **1999**, *125*, 195–202.
5. Cox, D.T.; Tissot, P.; Michaud, P. Water level observations and short-term predictions including meteorological events for entrance of Galveston Bay, Texas. *J. Waterway Port Coast. Ocean Eng.* **2002**, *128*, 1–29.
6. Kim, D.H.; Park, W.S. Neural network for design and reliability analysis of rubble mound breakwaters. *Ocean Eng.* **2005**, *32*, 1332–1349.
7. Van Gent, M.R.A.; Van den Boogaard, H.F.P.; Pozueta, B.; Medina, J.R. Neural network modeling of wave overtopping at coastal structures. *Coast. Eng.* **2007**, *54*, 586–593.
8. Browne, M.; Castelle, B.; Strauss, D.; Tomlinson, R.; Blumenstein, M.; Lane, C. Near-shore swell estimation from a global wind–wave model: Spectral process, linear and artificial neural network models. *Coast. Eng.* **2007**, *54*, 445–460.
9. Balas, C.E.; Koc, M.L.; Tür, R. Artificial neural networks based on principal component analysis, fuzzy systems and fuzzy neural networks for preliminary design of rubble mound breakwaters. *Appl. Ocean Res.* **2010**, *32*, 425–433.
10. Yoon, H.D.; Cox, D.T.; Kim, M.K. Prediction of time-dependent sediment suspension in the surf zone using artificial neural network. *Coast. Eng.* **2013**, *71*, 78–86.
11. Rabunal, J.R.; Dorado, J. *Artificial Neural Networks in Real-Life Applications*; Idea Group Publishing: Hershey, PA, USA, 2006.
12. Haykin, S. *Neural Networks: A Comprehensive Foundation*; Prentice Hall: Upper Saddle River, NJ, USA, 1999.
13. Rocha, M.; Cortez, P.; Neves, J. Evolutionary neural network learning. In *Progress in Artificial Intelligence*; Pires, F.M., Abreu, S., Eds.; Springer Berlin Heidelberg: Heidelberg, Germany, 2003; Volume 2902, pp. 24–28.
14. Krenker, A.; Bester, J.; Kos, A. Introduction to the artificial neural networks. In *Artificial Neural Networks—Methodological Advances and Biomedical Applications*; Suzuki, K., Ed.; InTech: Rijeka, Croatia, 2011; pp. 15–30.
15. Kolen, J.F.; Pollack, J.B. Back Propagation Is Sensitive to Initial Conditions. In Proceedings of the Advances in Neural Information Processing Systems 3, Denver, CO, USA, 26–29 November 1990; Lippmann, R.P., Moody, J.E., Touretzky, D.S., Eds.; Morgan Kaufmann: San Francisco, CA, USA, 1990; pp. 860–867.

16. Yam, Y.F.; Chow, T.W.S. Determining initial weights of feedforward neural networks based on least-squares method. *Neural Process. Lett.* **1995**, *2*, 13–17.

17. Venkatesan, D.; Kannan, K.; Saravanan, R. A genetic algorithm-based artificial neural network model for the optimization of machining processes. *Neural Comput. Appl.* **2009**, *18*, 135–140.

18. Chang, Y.-T.; Lin, J.; Shieh, J.-S.; Abbod, M.F. Optimization the initial weights of artificial neural networks via genetic algorithm applied to hip bone fracture prediction. *Adv. Fuzzy Syst.* **2012**, *2012*, 951247.

19. Mulia, I.E.; Tay, H.; Roopsekhar, K.; Tkalich, P. Hybrid ANN-GA model for predicting turbidity and chlorophyll-a concentration. *J. Hydroenv. Res.* **2013**, *7*, 279–299.

20. Krogh, A.; Vedelsby, J. Neural network ensembles, cross validation and active learning. *Adv. Neural Inf. Process. Syst.* **1995**, *7*, 231–238.

21. Boucher, M.A.; Perreault, L.; Anctil, F. Tools for the assessment of hydrological ensemble forecasts obtained by neural networks. *J. Hydroinf.* **2009**, *11*, 297–307.

22. Zamani, A.; Azimian, A.; Heemink, A.; Solomatine, D. Wave height prediction at the Caspian Sea using a data-driven model and ensemble-based data assimilation methods. *J. Hydroinf.* **2009**, *11*, 154–164.

23. Kim, S.E. Improving the Generalization Accuracy of ANN Modeling Using Factor Analysis and Cluster Analysis: Its Application to Streamflow and Water Quality Predictions. Ph.D. Thesis, Seoul National University, Seoul, Korea, 2014.

24. Geem, Z.W.; Kim, J.H.; Loganathan, G.V. A new heuristic optimization algorithm: Harmony search. *Simulation* **2001**, *76*, 60–68.

25. Hudson, R.Y. Laboratory investigation of rubble-mound breakwaters. *J. Waterways Harbors Div.* **1959**, *85*, 93–121.

26. Van der Meer, J.W. *Rock Slopes and Gravel Beaches under Wave Attack*; Delft Hydraulics Communication No. 396: Delft, The Netherlands, 1988.

27. Van der Meer, J.W. Stability of breakwater armor layers—Design formulae. *Coast. Eng.* **1987**, *11*, 93–121.

28. Smith, W.G.; Kobayashi, N.; Kaku, S. Profile Changes of Rock Slopes by Irregular Waves. In Proceedings of the 23rd International Conference on Coastal Engineering, Venice, Italy, 4–9 October 1992; Edge, B.L., Ed.; American Society of Civil Engineers: Reston, VA, USA, 1992; pp. 1559–1572.

29. Haldar, A.; Mahadevan, S. *Reliability Assessment Using Stochastic Finite Element Analysis*; John Wiley & Sons: New York, NY, USA, 2000.

30. Hoel, P.G. *Introduction to Mathematical Statistics*, 3rd ed.; Wiley & Sons: New York, NY, USA, 1962.

31. Lee, A. Determination of Near-Global Optimal Initial Weights of Artificial Neural Networks Using Harmony Search Algorithm: Application to Breakwater Armor Stones. Master's Thesis, Seoul National University, Seoul, Korea, 2016.

32. Lee, K.S.; Geem, Z.W. A new structural optimization method based on the harmony search algorithm. *Comput. Struct.* **2004**, *82*, 781–798.

33. Geem, Z.W. *Music-Inspired Harmony Search Algorithm*; Springer: Berlin, Germany, 2009.

34. Willmott, C.J. On the validation of models. *Phys. Geol.* **1981**, *2*, 184–194.

35. Whitley, D.; Starkweather, T.; Bogart, C. Genetic algorithms and neural networks: Optimizing connections and connectivity. *Parallel Comput.* **1990**, *14*, 347–361.

36. Hozjan, T.; Turk, G.; Fister, I. Hybrid artificial neural network for fire analysis of steel frames. In *Adaptation and Hybridization in Computational Intelligence*; Fister, I., Ed.; Springer International Publishing: Cham, Switzerland, 2015; pp. 149–169.

37. Zhang, J.R.; Zhang, J.; Lok, T.; Lyu, M.R. A hybrid particle swarm optimization–back-propagation algorithm for feedforward neural network training. *Appl. Math. Comput.* **2007**, *185*, 1026–1037.

38. Nikelshpur, D.; Tappert, C. Using Particle Swarm Optimization to Pre-Train Artificial Neural Networks: Selecting Initial Training Weights for Feed-Forward Back-Propagation Neural Networks. In Proceedings of the Student-Faculty Research Day, CSIS, Pace University, New York, NY, USA, 3 May 2013.

39. Nawi, N.M.; Khan, A.; Rehman, M.Z. A New Back-Propagation Neural Network Optimized with Cuckoo Search Algorithm. In *Computational Science and Its Applications–ICCSA 2013*, Proceedings of the 13th International Conference on Computational Science and Its Applications, Ho Chi Minh City, Vietnam, 24–27 June 2013; Part I. Murgante, B., Misra, S., Carlini, M., Torre, C., Nguyen, H.-Q., Taniar, D., Apduhan, B.O., Gervasi, O., Eds.; Springer Berlin Heidelberg: Heidelberg, Germany, 2013; Volume 7971, pp. 413–426.

Network Modeling and Assessment of Ecosystem Health by a Multi-Population Swarm Optimized Neural Network Ensemble

Rong Shan, Zeng-Shun Zhao, Pan-Fei Chen, Wei-Jian Liu, Shu-Yi Xiao, Yu-Han Hou, Mao-Yong Cao, Fa-Liang Chang and Zhigang Wang

Abstract: Society is more and more interested in developing mathematical models to assess and forecast the environmental and biological health conditions of our planet. However, most existing models cannot determine the long-range impacts of potential policies without considering the complex global factors and their cross effects in biological systems. In this paper, the Markov property and Neural Network Ensemble (NNE) are utilized to construct an estimated matrix that combines the interaction of the different local factors. With such an estimation matrix, we could obtain estimated variables that could reflect the global influence. The ensemble weights are trained by multiple population algorithms. Our prediction could fit the real trend of the two predicted measures, namely Morbidity Rate and Gross Domestic Product (GDP). It could be an effective method of reflecting the relationship between input factors and predicted measures of the health of ecosystems. The method can perform a sensitivity analysis, which could help determine the critical factors that could be adjusted to move the ecosystem in a sustainable direction.

Reprinted from *Appl. Sci.* Cite as: Shan, R.; Zhao, Z.-S.; Chen, P.-F.; Liu, W.-J.; Xiao, S.-Y.; Hou, Y.-H.; Cao, M.-Y.; Chang, F.-L.; Wang, Z. Network Modeling and Assessment of Ecosystem Health by a Multi-Population Swarm Optimized Neural Network Ensemble. *Appl. Sci.* **2016**, *6*, 175.

1. Introduction

Historical ecology research is increasingly valuable in assessing long-term baselines and understanding long-term ecological changes, and is increasingly applicable to management, decision-making, and conservation. When ecological survey data are lacking, historical data can be exploited to build mathematical models to supply scientifically meaningful information despite limitations in precision.

Soaring demands for food, fresh water, fuel, and timber have contributed to dramatic environmental changes. Nearly two-thirds of Earth's life-supporting ecosystems—including clean water, pure air, and a stable climate—are being degraded by unsustainable use. Many scientific studies have come to the conclusion that there is growing stress on Earth's biological systems. More and more warning signs are appearing. Is Planet Earth truly nearing a global tipping point? Is such an extreme state inevitable? Very few global models are proposed to address those claims and questions. As a result, society is more and more interested in developing mathematical models to assess and forecast the environmental and biological health conditions of our planet.

In Reference [1], the article presents two specific quantitative modeling challenges in their call for better predictive models:

(1) To improve bio-forecasting through global models that embrace the complexity of Earth's interrelated systems and include the effects of local conditions on the global system and *vice versa*.

(2) To identify factors that could produce unhealthy global state shifts and to show how to use effective ecosystem management to prevent or limit these impending state changes.

Determining whether global models can be built using local or regional components of the Earth's health that predict potential state changes is a huge challenge. Then, based on the potential impact on Earth's health, how can the global models help decision-makers design effective policies?

The "Nature" article [1] and many others point out that there are several important elements at work in the Earth's ecosystem (e.g., local factors, global impacts, multi-dimensional factors and relationships, varying time and spatial scales). There are also many other factors that can be included in a predictive model—human population, resource and habitat stress, habitat transformation, energy consumption, climate change, land use patterns, pollution, atmospheric chemistry, ocean chemistry, biodiversity, and political patterns such as social unrest and economic instability. Paleontologists have studied and modeled ecosystem behavior and response during previous cataclysmic state shifts and thus historic-based qualitative and quantitative information can provide background for future predictive models. However, it should be noted that human effects have increased significantly in our current biosphere situation.

Reference [2] introduced the "WORLD3" model, which is based on system dynamics—a method for studying the world that deals with understanding how complex systems change over time. Internal feedback loops within the structure of the system influence the entire system's behavior. However, the model does not make the prediction; rather, it is a tool for understanding the broad sweeps and the behavioral tendencies of the system.

Scientists realize that it is very important to assess and predict the potential state changes of the planetary health systems. Nowadays, there is considerable research [3–8] being conducted that takes local habitats and regional factors into account. However, since the article [1] published in "Nature" called for better predictive models in 2012, there have been few models addressing the problem of predicting long-range and global impacts based only on local factors. Using current models, decision-makers cannot be adequately informed about how their provincial polices may impact the overall health of the planet. Many existing models cannot determine the long-range impacts of potential policies without considering the complex global factors, the complex relationships, and the cross effects in biological systems. The system complexities are manifested in multiple interactions, feedback loops, emergent behaviors, and impending state changes or tipping points. As is well known, suggesting suitable potential policies is very important for the sustainable ecological development of our planet. However, the first step should be to determine those critical factors that affect the global state. Based on the idea above, this paper aims at proposing a framework to model and assess the interactive roles of the local factors, to determine the critical factors and to suggest potential management for the ecosystem's health.

2. Problem Definition and Model Design

2.1. Defining the Problem

With the rapid economic development, the level of energy demand is rapidly increasing, which results in a series of environmental problems. Both urban and rural ecosystems have to carry out environmental performance evaluation in energy utilization to reconcile economic growth with ecological preservation. We focus on the following considerations:

(1)   How to construct a dynamic global network model that includes dynamic elements to predict future states of ecological health.
(2)   How to determine the critical factors that reflect the relationship between the model and the predictive measure.
(3)   How to determine a feedback policy that reflects the influence of human factors.

2.2. Basic Assumptions

There are so many elements that could be included in the ecological modeling. For simplicity, we make the following assumptions:

(1)   The state changes over the years and we can observe these state changes.
(2)   The $k - th$ factor at the t moment is influenced by the same factor at the $t - 1$ moment and interactive factors in neighborhood at the t moment; we ignore the influence between the different factors at the $t - 1$ moment. There has been Markov property.
(3)   The relative influence between one factor and another factor just reflects the ratio of two value changes.
(4)   To simplify the model, we select several typical factors.

(5) Outside factors such as "universe perishing", a destructive earthquake, and a volcanic eruption will not be considered.

(6) Each factor has affected other factors and vice versa. The influence is regarded as directive.

(7) The CO_2 content could reflect environmental changes.

(8) The oil consumption could reflect the source consumption.

(9) The population would reflect the bearing capacity of the ecosystem.

(10) The electricity would reflect the power consumption.

(11) The morbidity rate would reflect threats to life.

(12) The Gross Domestic Product (GDP) would reflect the level of wellbeing.

2.3. Symbol Lists

In the concrete implementation of the proposed framework, only four typical factors are utilized in the ecological modeling, as denoted in Table 1 and Figure 1. As we adopt the Markov model to predict the dynamic elements, there are also four corresponding symbols to represent their respective predictions.

Table 1. The symbol lists.

Variables	Items in Figure 1	Meaning of Variables
$\alpha(t)$	Local input factor 1 at the t moment	The CO_2 Content
$\hat{\alpha}(t)$	Estimated input factor 1 at the t moment	The estimated CO_2 content
$\beta(t)$	Local input factor 2 at the t moment	The oil consumption
$\hat{\beta}(t)$	Estimated input factor 2 at the t moment	The estimated oil consumption
$\gamma(t)$	Local input factor 3 at the t moment	The population
$\hat{\gamma}(t)$	Estimated input factor 3 at the t moment	The estimated population
$\delta(t)$	Local input factor 4 at the t moment	The power consumption
$\hat{\delta}(t)$	Estimated input factor 4 at the t moment	The estimated power consumption
$\phi(t)$	Predicted measurement 1 at the t moment	The morbidity rate
$\mu(t)$	Predicted measurement 2 at the t moment	The Gross Domestic Product (GDP)
$\Delta()$	The change between two continual times for each factor	The change between two continual times for each factor

2.4. Model Design

Our model is composed of three segments: the first estimates the global dynamic influence of factors via the local interacting factors. The second predicts the measured values for the ecosystem that could reflect the whole ecosystem's state. The third determines the critical factors and loop control to adjust the ecosystem. In Figure 1, we present the schematic diagram as follows.

In the colorful block, the white double arrow denotes the interacting influence between two local factors. From $t-1$ moment to t moment, it is a dynamic alignment that incorporates the Markov property. The flow arrows mean the prediction of the neural networks. A, B, and C denote the different government policies. The blue single side arrow denotes the policy acting on the input factors. The feedback circle means that the government takes measures to tackle the global state change.

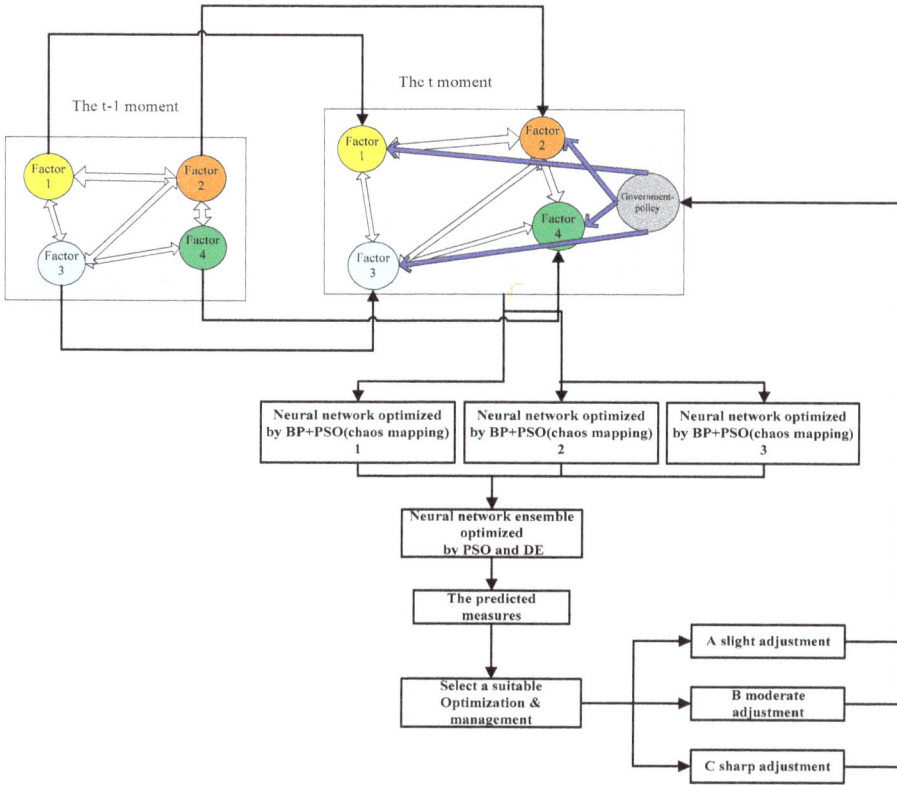

Figure 1. The model schematic diagram (for example: factor 1 denotes CO_2 content, factor 2 denotes oil consumption, factor 3 denotes population, and factor 4 denotes the power consumption).

3. Model Solution

3.1. Estimating the Dynamic Factors

In the first sub-section, we presented the Markov chain to reflect the influence between the t moment and $t-1$ moment. We assumed that the $k-th$ factor at the t moment is influenced by the same factor at the $t-1$ moment and the interacting factors in neighborhood at the t moment, and ignore the influence between different factors at the $t-1$ moment, which is reasonable in order to simplify the model by the Markov property. Thus we could construct an estimation matrix M that incorporates the interactive relationship of different dynamic factors. In Table 2, we list the estimated Markov impacting matrix M.

Table 2. The estimated Markov impacting matrix M.

Local Factors	α	β	γ	δ
α	$\frac{\alpha(t)-\alpha(t-1)}{\alpha(t)}+1$	$\frac{\gamma(t)-\gamma(t-1)}{a(t)-\alpha(t-1)}$	$\frac{\gamma(t)-\gamma(t-1)}{\alpha(t)-\alpha(t-1)}$	$\frac{\delta(t)-\delta(t-1)}{\alpha(t)-\alpha(t-1)}$
β	$\frac{\alpha(t)-\alpha(t-1)}{\beta(t)-\beta(t-1)}$	$\frac{\beta(t)-\beta(t-1)}{\beta(t)}+1$	$\frac{\gamma(t)-\gamma(t-1)}{\beta(t)-\beta(t-1)}$	$\frac{\delta(t)-\delta(t-1)}{\beta(t)-\beta(t-1)}$
γ	$\frac{\alpha(t)-\alpha(t-1)}{\gamma(t)-\gamma(t-1)}$	$\frac{\beta(t)-\beta(t-1)}{\gamma(t)-\gamma(t-1)}$	$\frac{\gamma(t)-\gamma(t-1)}{\gamma(t)}+1$	$\frac{\delta(t)-\delta(t-1)}{\gamma(t)-\gamma(t-1)}$
δ	$\frac{\alpha(t)-\alpha(t-1)}{\delta(t)-\delta(t-1)}$	$\frac{\beta(t)-\beta(t-1)}{\delta(t)-\delta(t-1)}$	$\frac{\gamma(t)-\gamma(t-1)}{\delta(t)-\delta(t-1)}$	$\frac{\delta(t)-\delta(t-1)}{\delta(t)}+1$

The elements in the leading diagonal denote a continuous effects on the same factor over time ($t-1$ & t). The other elements in the matrix present the effect on different factors at the same t moment. The difference between the $k-th$ factor and the $m-th$ factor reflects on characteristic variations of industrial and agricultural production, consumption, and populations at the same time period.

We could obtain the estimated dynamic factors using the following formula:

$$[\hat{\alpha}(t)\hat{\beta}(t)\hat{\gamma}(t)\hat{\delta}(t)] = [\alpha(t)\beta(t)\gamma(t)\delta(t)]\cdot \begin{bmatrix} M_{11} & M_{12} & M_{13} & M_{14} \\ M_{21} & M_{22} & M_{23} & M_{24} \\ M_{31} & M_{32} & M_{33} & M_{34} \\ M_{41} & M_{42} & M_{43} & M_{44} \end{bmatrix} \tag{1}$$

We regard $[\hat{\alpha}(t)\hat{\beta}(t)\hat{\gamma}(t)\hat{\delta}(t)]$ as the dynamic factors of the Neural Network Ensemble (NNE). Via all these input factors above, we could predict the output measure.

3.2. Predicting the Output Measure

Having obtained the estimated dynamic factors, the output measure of the ecological modeling can be predicted using our previously proposed Evolved Neural Network Ensemble (NNE), improved by Multiple Heterogeneous Swarm Intelligence [9]. Compared to the ordinary NNE, to improve the prediction precision, we incorporate the Particle Swarm Optimization (PSO) [10,11] and Back Propagation (BP) algorithms to train each component Forward Neural Network (FNN). Meanwhile, we apply logistic chaos mapping to enhance the local searching ability. At the same time, the ensemble weights are trained by multi-population algorithms (PSO and Differential Evolution (DE) [9] cooperative algorithms are used in this case). By the NNE algorithms, we could remove the disturbance in the data. A more detailed description is given in [9]. In Figure 2, we summarize the schematic diagram of the improved NNE as follows:

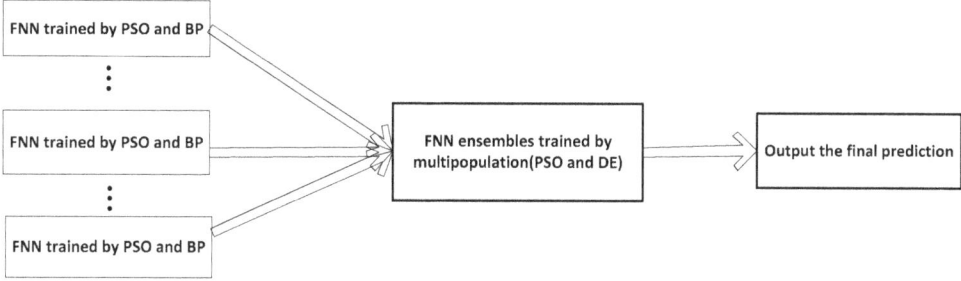

Figure 2. The Neural Network Ensemble (NNE) schematic diagram. FNN: Forward Neural Network; PSO: Particle Swarm Optimization; BP: Back Propagation; DE: Differential Evolution.

We could divide this sub-section into two parts: one is how to train and optimize the component neural networks; the other is how to optimize the NNE by multi-population algorithms.

3.2.1. Optimization by Particle Swarm Optimization (PSO) Algorithm, Chaotic Mapping, and Back Propagation (BP) Algorithm

The PSO could be described as a swarm of birds hovering in the sky for food. $X_i = (x_{i1}, x_{i2}, \ldots, x_{in})$ and $V_i = (v_{i1}, v_{i2}, \ldots, v_{in})$, where X_i and V_i are the position and velocity of the $i - th$ particle in n dimensional space. $P_i = (p_{i1}, p_{i2}, \ldots, p_{in})$ represents the previous best value of the $i - th$ particle up to the current step. $G_i = (g_{i1}, g_{i2}, \ldots, g_{in})$ represents the best value of all particles in the population. Having obtained the two best values, each particle updates its position and velocity according to the following equations:

$$v_{i,d}^{k+1} = w \cdot v_{i,d}^{k} + c_1 \cdot rand() \cdot \left(pbest_{i,d}^{k} - x_{i,d}^{k} \right) + c_2 \cdot rand() \cdot \left(gbest_{d}^{k} - x_{i,d}^{k} \right) \tag{2}$$

$$x_{i,d}^{k+1} = x_{i,d}^{k} + v_{i,d}^{k+1} \tag{3}$$

c_1 and c_2 are learning actors. w^t is the inertial weight. The flow chart depicting the general PSO Algorithm is given in Figure 3. A more detailed description is given in [9–11].

In addition, with the purpose of enhancing the local searching ability and diversity of the particle swarm, we incorporate the chaos mechanism [12] into the updating iteration. We take the logistic mapping as the chaos producer.

The well-known idea of BP is to make the error back propagate to update the parameters of FNN, and the parameters include two steps: one is between the input layer and hidden layer; the other is between the hidden layer and the output layer.

A detailed description of the procedures of the PSO (combined logistic mapping)–BP coupled algorithm is presented is in [9,10].

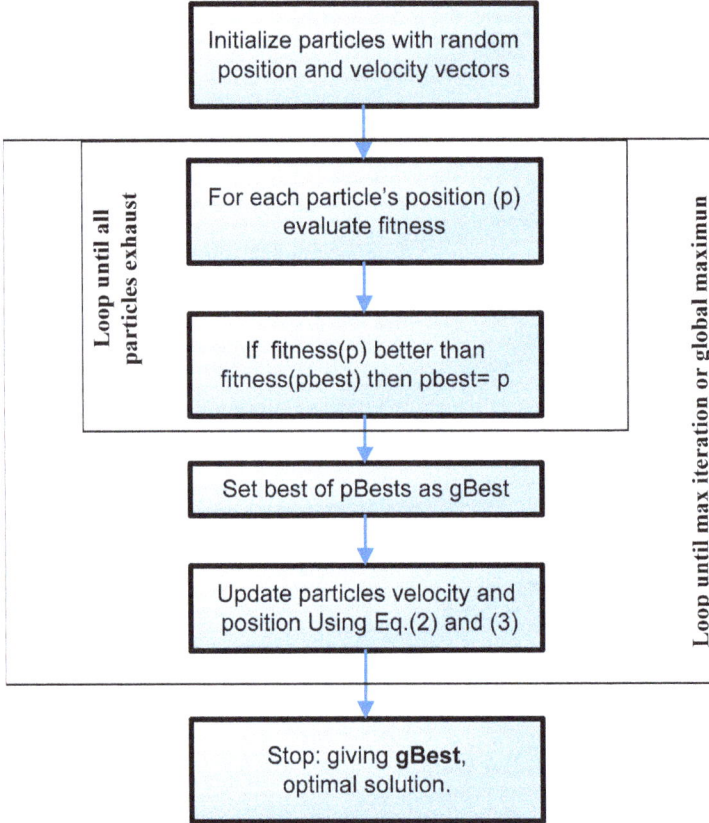

Figure 3. The flow chart of PSO algorithm.

3.2.2. Optimization by Multi-Population Cooperative Algorithm

The principle of the NNE has been described in detail in [9,13,14]. Having obtained each refined component FNN, we would concentrate on how to combine the output of each component FNN.

$$\overline{f}(x) = \sum_{i=1}^{n} \overline{w}_i f_i(x) \tag{4}$$

$$\overline{w}_i = \frac{w_i}{\sum_{i=1}^{n} w_i}, 0 < w_i < 1 \tag{5}$$

where $f_i(x)$ represents the output of the $i - th$ FNN and $\overline{w}_i$ represents the importance of the $i - th$ FNN. Our idea is to obtain the most appropriate prediction of $\overline{w}_i$ for each sub-network, which corresponds to the solution of the optimization problem of the particles. DE is also a floating, point-encoded evolutionary algorithm for global optimization over continuous spaces, but it creates new candidate solutions by combining the parent individual and several other individuals of the same population. It consists of selection, crossover, and mutation [15].

The procedure for NNE multi-population algorithm could be summarized as follows:

Step 1: Initialize the weight of each FNN, which has been optimized by PSO (combined logistic mapping) and BP.

Step 2: Each particle represents a set of weights, which means that each dimension represents one weight of each component FNN. The population is duplicated into two identical swarms.

Step 3: One swarm is optimized by PSO and the other is optimized by DE.

Step 4: Update the $gbest\_PSO$ and $gbest\_DE$.

Step 5: Do the Steps 3–4 loop until the Max-iteration is reached.

Step 6: Output the predicted measures.

3.3. Optimization and Management

In the third sub-section, we could determine the critical factors by sensitivity analysis and determine the measurement standards that could evaluate the state change. The local state change could influence the global state and vice versa. Via a different management mechanism, government policy could be established and the ecosystem could be developed according to the sustainable direction.

3.3.1. Sensitivity Design

Having obtained the dynamic predicted measures, we could calculate the sensitivity between the input factors and the predicted measures.

We determined $\Delta(t)$, which denotes a change of input or output state, and obtained the values using the following formulae:

$$\Delta \hat{\alpha}(t) = \hat{\alpha}(t) - \hat{\alpha}(t-1) \tag{6}$$

$$\Delta \hat{\beta}(t) = \hat{\beta}(t) - \hat{\beta}(t-1) \tag{7}$$

$$\Delta \hat{\gamma}(t) = \hat{\gamma}(t) - \hat{\gamma}(t-1) \tag{8}$$

$$\Delta \hat{\delta}(t) = \hat{\delta}(t) - \hat{\delta}(t-1) \tag{9}$$

$$\Delta \varphi(t) = \varphi(t) - \varphi(t-1) \tag{10}$$

$$\Delta\mu(t) = \mu(t) - \mu(t-1). \tag{11}$$

We could calculate the sensitivity between the predicted measure and the estimated input factors using the following formula:

$S(y, x) = \frac{dy}{dx} \times \frac{x}{y}$, which could take the approximated format:

$$S(y, x) = \frac{\Delta y}{\Delta x} \times \frac{x}{y} \tag{12}$$

We would get the following results:

$$S(\varphi(t), \alpha(t)) = \frac{\Delta\varphi(t)}{\Delta\alpha(t)} \times \frac{\alpha(t)}{\varphi(t)} \tag{13}$$

$$S(\varphi(t), \beta(t)) = \frac{\Delta\varphi(t)}{\Delta\beta(t)} \times \frac{\beta(t)}{\varphi(t)} \tag{14}$$

$$S(\varphi(t), \gamma(t)) = \frac{\Delta\varphi(t)}{\Delta\gamma(t)} \times \frac{\gamma(t)}{\varphi(t)} \tag{15}$$

$$S(\varphi(t), \delta(t)) = \frac{\Delta\varphi(t)}{\Delta\delta(t)} \times \frac{\delta(t)}{\varphi(t)} \tag{16}$$

$$S(\mu(t), \alpha(t)) = \frac{\Delta\mu(t)}{\Delta\alpha(t)} \times \frac{\alpha(t)}{\mu(t)} \tag{17}$$

$$S(\mu(t), \beta(t)) = \frac{\Delta\mu(t)}{\Delta\beta(t)} \times \frac{\beta(t)}{\mu(t)} \tag{18}$$

$$S(\mu(t), \gamma(t)) = \frac{\Delta\mu(t)}{\Delta\gamma(t)} \times \frac{\gamma(t)}{\mu(t)} \tag{19}$$

$$S(\mu(t), \delta(t)) = \frac{\Delta\mu(t)}{\Delta\delta(t)} \times \frac{\delta(t)}{\mu(t)}. \tag{20}$$

If $S(\cdot, \cdot) < 0$, it denotes a negative influence. The predicted measures would decrease, with the input factor increasing.

If $S(\cdot, \cdot) > 0$, it means a positive influence. With the input factor increasing, the predicted measures would increase.

The bigger the absolute value of $S(\cdot, \cdot)$ is, the more powerful the influence between the input factor and the predicted measures would be. By this means we could obtain the critical factor, which acts on the global influence.

3.3.2. Management and Policy

We take the fuzzy rules to determine the measure standard, and we also take the last 5th year as the comparison. With several values as thresholds, the rules can be described as follows:

If $\frac{\varphi(t)-\varphi(t-1)}{\varphi(t)-\varphi(t-5)} > 3\%$, it means that the state has changed slightly. We could adopt policy A, which acts on the critical factor of φ.

Otherwise, if $\frac{\varphi(t)-\varphi(t-1)}{\varphi(t)-\varphi(t-5)} > 5\%$, it means that the state has changed moderately. Alternatively, we could adopt policy B, which acts on the critical factor of φ.

Otherwise, if $\frac{\varphi(t)-\varphi(t-1)}{\varphi(t)-\varphi(t-5)} > 8\%$, it means that the state has seriously changed. Lastly, we could adopt policy C, which acts on the critical factor of φ.

End If.

In Figure 4, we present the management diagram as follows:

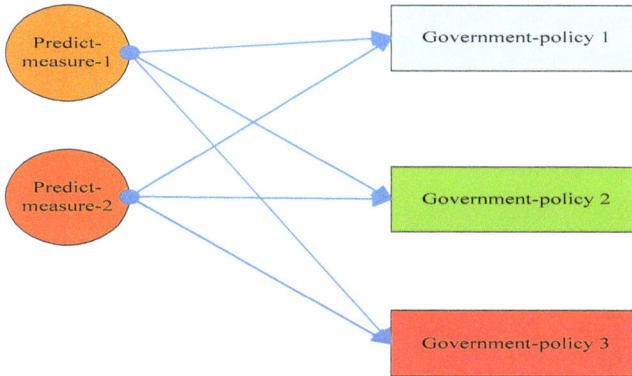

Figure 4. The management diagram.

We choose the suitable policy to control the critical factor of each predicted value. We also make a rule table as follows. By this means we could obtain a suitable policy for overall prediction of φ, μ. In Table 3, we give the rules table.

Table 3. The rules table.

Items	Values							
Predicted measurement-1	0	A	A	B	B	A	B	C
Predicted measurement-2	0	0	A	A	B	C	C	C
The final policy	0	A	A	B	B	C	C	C

We could make a brief policy mechanism with the pseudo-code description given in Table 4.

51

Table 4. The pseudo-code description of the decision-making strategy.

For the two predicted values:
    If there are more than one policy Cs
        We would apply policy C to all the input factors.
    If there are more than one policy Bs
        We would apply policy B to all the input-factors.
    If there are more than one policy As
        We would apply policy A to all the input-factors.
    Otherwise do not intervene.
End If.

Based on the predicted measures, we could suggest that the government draft detailed policies. For example:

Policy A: Increase public awareness of (1) the damage caused by increasing CO_2; (2) the damage caused by logging; (3) the dangers linked to an increasing population; and (4) strategies for economizing on electricity.

Policy B: Limit (1) car exhaust; (2) large-scale logging; (3) population by slight measures; (4) electrical consumption in certain areas or certain time periods; and (5) develop novel technologies to tackle the problem.

Policy C: (1) Make laws to limit CO_2 and fuel consumption and (2) advocate forest planting.

4. Data Testing and Analysis

4.1. Data Design

The dataset for ecological modeling is collected from references [8,16–18] (unfortunately, we cannot access all the fields of the dataset for 2007–2015). The local factor data and the real measured data are shown in the Tables 5 and 6, respectively.

Table 5. The local factor data. 1e3 means ten to the power of 3.

Year	CO₂ (1e3 Billion Ton)	Oil Consumption	Population (10 Billion)	Power Consumption (1e5 $kw\ h$)
1996	0.36236	0.429	0.57962	0.136714
1997	0.36347	0.445	0.5878	0.139623
1998	0.3665	0.455	0.5959	0.143186
1999	0.36814	0.46	0.60386	0.147216
2000	0.3694	0.475	0.61181	0.154169
2001	0.37107	0.48	0.61957	0.155235
2002	0.37317	0.482	0.62725	0.161266
2003	0.37578	0.483	0.63492	0.167138
2004	0.37752	0.49	0.64261	0.175049
2005	0.37976	0.51	0.65032	0.182721
2006	0.38185	0.515	0.65805	0.189788

Table 6. The real, measured data.

	The Real, Measured Data	
Year	Morbidity Rate (100%)	GDP (10 per Unit)
1996	0.406	0.4520
1997	0.411	0.4660
1998	0.438	0.4750
1999	0.454	0.4811
2000	0.442	0.4950
2001	0.437	0.5031
2002	0.435	0.5061
2003	0.433	0.5062
2004	0.411	0.5063
2005	0.407	0.5151
2006	0.405	0.5280

Factors such as CO_2 content, oil consumption, population, and power consumption are regarded as the input factors. Using the Markov estimation, we could get the estimated input factors.

4.2. Training and Analysis

We adopt the data from 1996 to 2001 to train our neural networks. The input variables include CO_2, oil consumption, population, and power consumption, and the predicted measures include the morbidity rate and GDP. A sensitivity analysis is conducted to determine the critical factors and put forward advice that would help the government to establish policy. In Table 7, we list the sensitivity of the input factors and predicted measures from 1997 to 2001.

From Table 7, we could draw a conclusion that the CO_2 content has been the critical factor. In our model, CO_2 stands for the environmental change.

In 1997, CO_2 was the main factor behind the rapid increase of GDP, and also strengthened the morbidity rate. This means that the GDP increasing is based on environmental deterioration. Meanwhile, the deteriorating environment has caused an increase in the morbidity rate.

From 1998 to 1999, CO_2 has also been the prominent factor in the rapid increasing of GDP, and exacerbated the morbidity rate. However, the increase has plateaued, which means that the government has recognized the problem and taken slight measures to tackle it.

In 2000, CO_2 shifted from a positive role to a negative factor in relation to the morbidity rate, but meanwhile it strengthened the GDP. This means that the government has taken measures to prevent some diseases. However, GDP increasing is also based on environmental deterioration.

Table 7. The sensitivity of the local factors and predicted measures. The bold words show the maxmimum absolute value about the sensitivity analysis.

Analysis of Sensitivity		CO_2 Content	Oil Consumption	Population	Power Consumption
1997	Morbidity rate	**3.9836**	0.3384	0.8742	0.5839
	The GDP	**9.8376**	0.8305	2.1947	1.4557
1998	Morbidity rate	**7.4563**	2.8048	4.5350	2.4771
	The GDP	**2.2918**	0.8621	1.3939	0.7614
1999	Morbidity rate	**7.9110**	3.2423	2.6735	1.2875
	The GDP	**2.8001**	1.1476	0.9463	0.4557
2000	Morbidity rate	**−7.9595**	−0.8597	−2.0893	−0.6020
	The GDP	**8.2918**	0.8956	2.2108	0.6163
2001	Morbidity rate	**−2.5423**	−1.0984	−0.9135	−1.6662
	The GDP	**−3.5340**	1.5268	1.2698	2.3161

In 2001, CO_2 was a negative factor in relation to the GDP and morbidity rate. By a sensitivity analysis of morbidity rate, we found that environmental deterioration has alleviated the positive influence of the government policies that prevent some diseases. Even more, the environmental deterioration has hindered the increasing of GDP.

4.3. Testing and Analysis

We use data from 2001–2006 to test our model. We could draw the conclusion that our model could fit the approximating trend of the two predicted measures. In Table 8, we list the sensitivity of the input factors and predicted measures from 2002 to 2006.

Table 8. The sensitivity of the local factors and predicted measures of the test data. The bold words show the maxmimum absolute value about the sensitivity analysis.

Analysis of Sensitivity		CO_2	Oil Consumption	Population	Power Consumption
2002	Morbidity rate	−0.8170	**−1.1080**	−0.3755	−0.1229
	GDP	1.0885	**1.4762**	0.5003	0.1638
2003	Morbidity rate	−0.6650	**−2.2309**	−0.3824	−0.1315
	GDP	0.0284	**0.0954**	0.0164	0.0056
2004	Morbidity rate	**−11.6137**	−3.7470	−4.4730	−1.1744
	GDP	**0.0429**	0.0138	0.0165	0.0044
2005	Morbidity rate	**−1.6662**	−0.2506	−0.8290	−0.2341
	GDP	**2.8640**	0.4308	1.4249	0.4023
2006	Morbidity rate	**−0.9022**	−0.5086	−0.4204	−0.1326
	GDP	**4.4984**	2.5360	2.0960	0.6612

From Table 8, we could draw a conclusion that in 2002–2003 the oil consumption was the critical factor and in 2004–2006 CO_2 content was the critical factor. In our model, CO_2 denotes environmental change and oil consumption could represent resource consumption.

In 2002, oil consumption was a positive factor in relation to the rapid increasing of GDP, but strengthened the morbidity rate. In the last five years, the government has adopted several policies to tackle environmental deterioration. However, it must be noticed that GDP increasing is also based on resource consumption. In the meantime, excessive resource consumption has brought about an increase in the morbidity rate. The system must be monitored for both intended and unintended changes. The government should take measures to optimize the resource consumption mechanism.

Compared with the sensitivity of 2002, we could see that excessive resource consumption has further strengthened the morbidity rate, and also limited the increase of GDP. However, by 2002, the government had not taken effective measures.

In 2004–2006, the morbidity rate was alleviated by government policy. The GDP improved from 2004 to 2006. The government adopted some beneficial environmental policies but, over time, overconsumption of resources always negatively affects the environment. For some accumulated reasons, the CO_2 content could not be decreased but could be controlled to some extent.

5. Conclusions and Discussion

Our predictive model could fit the real trends. It could be an effective method to reflect the relationship between input factors and predicted measurements; in addition, the model helps to determine critical factors and measurement standards. By referring to this model, the government could make a detailed policy for adjusting the ecosystem. However, it must be noticed that when a large number of factors are considered, we have to construct a complicated matrix to reflect the dynamic relationship; so the simulation would be heavy, so a prominent component analysis would be considered.

Our method has the following strengths:

Firstly, our model can incorporate the relationships between different local state factors, taking into account different changing variables.

Secondly, the Neural Network Ensemble (NNE) is used to predict the global state. The predicting of single neural networks would be sensitive to disturbance. However, NNE could improve the stability of the model. In addition, PSO with logistic chaotic mapping could optimize the parameters in the networks and improve precision. The multi-population cooperative algorithm could enhance the stability of the NNE.

Lastly, by the analysis of sensitivity, our model could confirm the critical factors that affect the global state. Moreover, our model could determine the measurement standards used to select a policy.

Acknowledgments: This work was supported in part by the National Natural Science Foundation of China (Grant No. 61403281), the Natural Science Foundation of Shandong Province (ZR2014FM002), and a China Postdoctoral Science Special Foundation Funded Project (2015T80717).

Author Contributions: Rong Shan wrote the source code, revised the article; Zeng-Shun Zhao designed the algorithm, wrote the manuscript, analysed and interpretated the data; Pan-Fei Chen and Wei-Jian Liu conceived and designed the experiments; Shu-Yi Xiao and Yu-Han Hou performed the experiments; Mao-Yong Cao, Fa-Liang Chang and Zhigang Wang analyzed the data.

Conflicts of Interest: The authors declare no conflict of interest.

References

1. Barnosky, A.D.; Hadly, E.A.; Bascompte, J.; Berlow, E.L.; Brown, J.H.; Fortelius, M.; Getz, W.M.; Harte, J.; Hastings, A.; Marquet, P.A.; *et al.* Approaching a state shift in Earth's biosphere. *Nature* **2012**, *486*.

2. Mills, J.I.; Emmi, P.C. Limits to growth: The 30-year update. *J. Policy Anal. Manag.* **2006**, *25*, 241–245.

3. Scheffer, M.; Bascompte, J.; Brock, W.A.; Brovkin, V.; Carpenter, S.R.; Dakos, V.; Held, H.; van Nes, E.H.; Rietkerk, M.; Sugihara, G. Early-warning signals for critical transitions. *Nature* **2009**, *461*, 53–59.

4. Drake, J.M.; Griffen, B.D. Early warning signals of extinction in deteriorating environments. *Nature* **2010**, *467*, 456–459.

5. Brown, J.H.; Burnside, W.R.; Davidson, A.D.; DeLong, J.P.; Dunn, W.C.; Hamilton, M.J.; Mercado-Silva, N.; Nekola, J.C.; Okie, J.G.; Woodruff, W.H.; *et al.* Energetic limits to economic growth. *Bioscience* **2011**, *61*, 19–26.

6. McDaniel, C.N.; Borton, D.N. Increased human energy use causes biological diversity loss and undermines prospects for sustainability. *Bioscience* **2002**, *52*, 929–936.

7. Maurer, B.A. Relating human population growth to the loss of biodiversity. *Biodivers. Lett.* **1996**, *3*, 1–5.

8. Lavergne, S.; Mouquet, N.; Thuiller, W.; Ronce, O. Biodiversity and climate change: Integrating evolutionary and ecological responses of species and communities. *Annu. Rev. Ecol. Evol. Syst.* **2010**, *41*, 321–350.

9. Zhao, Z.; Feng, X.; Lin, Y.; Wei, F.; Wang, S.; Xiao, T.; Cao, M.; Hou, Z. Evolved neural network ensemble by multiple heterogeneous swarm intelligence. *Neuro Comput.* **2015**, *149*, 29–38.

10. Zhao, Z.; Feng, X.; Lin, Y.; Wei, F.; Wang, S.; Xiao, T.; Cao, M.; Hou, Z.; Tan, M. Improved Rao-Blackwellized particle filter by particle swarm optimization. *J. Appl. Math.* **2013**, *2013*, 1–7.

11. Kennedy, J.; Eberhart, R. Particle Swarm Optimization. In Proceedings of the IEEE Conference on Neural Networks, Piscataway, NJ, USA, 27 November 1995; pp. 1942–1948.

12. Gulick, D. *Encounters with Chaos*; McGraw Hill, Inc.: New York, NY, USA, 1992; pp. 127–186, 195–220, 240–285.

13. Optiz, D.; Shavlik, J. Actively searching for an effectively neural network ensemble. *Connect. Sci.* **1996**, *8*, 337–353.
14. Valentini, G.; Masulli, F. Ensembles of learning machines. *WIRN VIETRI* **2002**, *2002*, 3–20.
15. Storn, R.; Price, K. Differential evolution—A simple and efficient heuristic for global optimization over continuous spaces. *J. Glob. Optim.* **1997**, *11*, 341–359.
16. Data-The World Bank. Available online: http://data.worldbank.org.cn/ (accessed on 12 December 2015).
17. US-The Census Bureau. Available online: http://www.census.gov/compendia/statab/cats/births_deaths_marriages_divorces.html (accessed on 12 December 2015).
18. Population-Polution. Available online: http://www.google.com.hk/publicdata/explore?ds=d5bncppjof8f9_&met_y=sp_pop_totl&tdim=true&dl=zh-CN&hl=zh-CN&q=%E5%85%A8%E7%90%83%E4%BA%BA%E5%8F%A3 (accessed on 20 November 2015).

Simulation of Reservoir Sediment Flushing of the Three Gorges Reservoir Using an Artificial Neural Network

Xueying Li, Jun Qiu, Qianqian Shang and Fangfang Li

Abstract: Reservoir sedimentation and its effect on the environment are the most serious world-wide problems in water resources development and utilization today. As one of the largest water conservancy projects, the Three Gorges Reservoir (TGR) has been controversial since its demonstration period, and sedimentation is the major concern. Due to the complex physical mechanisms of water and sediment transport, this study adopts the Error Back Propagation Training Artificial Neural Network (BP-ANN) to analyze the relationship between the sediment flushing efficiency of the TGR and its influencing factors. The factors are determined by the analysis on 1D unsteady flow and sediment mathematical model, mainly including reservoir inflow, incoming sediment concentration, reservoir water level, and reservoir release. Considering the distinguishing features of reservoir sediment delivery in different seasons, the monthly average data from 2003, when the TGR was put into operation, to 2011 are used to train, validate, and test the BP-ANN model. The results indicate that, although the sample space is quite limited, the whole sediment delivery process can be schematized by the established BP-ANN model, which can be used to help sediment flushing and thus decrease the reservoir sedimentation.

Reprinted from *Appl. Sci.* Cite as: Li, X.; Qiu, J.; Shang, Q.; Li, F. Simulation of Reservoir Sediment Flushing of the Three Gorges Reservoir Using an Artificial Neural Network. *Appl. Sci.* **2016**, *6*, 148.

1. Introduction

Building reservoirs on rivers, especially sandy rivers, breaks the natural equilibrium state of flow and sediment conditions, as well as riverbed morphology. The lifting of the water level increases water depth, slows down the current velocity, and thus reduces the sediment carrying capacity of water, leading to a large number of sediment deposits in the reservoir.

The global capacity loss of reservoirs takes up to 0.5%–1% of the total reservoir storage every year, approximately 45 km^3 [1]. In addition to capacity loss, which disables the design function of reservoirs such as flood control, power generation, irrigation, and water supply, reservoir sedimentation also shortens the serve life of the reservoir, enlarges the flooded and submerged area upstream, threatens the safety

58

upstream, and impacts navigation. The reduction of outflow sediment also results in erosion of riverbeds downstream, and brings about a series of new problems.

There are plenty of sandy rivers in China. According to incomplete statistics, the annual sediment runoff of 11 main rivers in China is about 16.9×10^8 t [2]. The annual sediment runoff at the Cuntan hydrometric station of the Yangtze River is up to 2.10×10^8 t in the year 2012 [3]. With the construction of more and more large-scale reservoirs, the sedimentation problem has become prominent. The world-renowned Three Gorges Reservoir (TGR) is large multi-objective comprehensive water control project with functions of flood control, power generation, navigation, water supply, and so on. Ever since its design and demonstration phase, the sedimentation problem has been one of the principal concerns. Particle size of both the inflow and outflow sediment of the TGR presents a decreasing tendency, especially after the construction of TGR as well as other large-scale reservoirs upstream. The median particle diameter of the inflow sediment into the TGR is around 0.01–0.011 mm, while that of the outflow sediment is about 0.007–0.009 mm [3]. More than 85% of the inflow sediment particles are within 0.062 mm, and only 3%–7% are larger than 0.125mm. Thus, flushing is believed to be an effective way to transport sediment.

In order to make better use of reservoir release to flush sediment, the relationship between the amount of flushed sediment and its influence factors needs to be studied. However, sedimentation affected by geographical location, topography, geology, climate, and other natural factors, as well as human activities, is a very complex and cross-disciplinary subject. It involves river dynamics, geology, geography and other subjects with immature development. Both the complexity and the immaturity lead to the difficulties of solving the reservoir sedimentation problem. Calculation of sediment erosion and deposition can adopt the mathematical model of flow and sediment. There have been a range of mathematical models presented in previous publications [4–9], the theoretical foundation of which are mainly composed of: flow continuity equation, flow motion equation, sediment continuity equation, riverbed deformation equation, and sediment-carrying capacity equation. Establishing and solving a comprehensive set of equations for flow and sediment mathematical models is a complex task with requirements of extensive data to get closed conditions. Currently, many important aspects of the sediment transportation rely on experience and subjective judgment, which is not conducive to close the mathematical model. Furthermore, reservoir sedimentation cannot be generalized with a simple 1D mathematical model, while the 2D and 3D models for sediment transport are only available for short river reach or partial fluvial process issues due to their complex structure, the large number of nodes, and time-consuming computation [10–12].

The artificial neural network (ANN) model has the characteristics of parallelism, robustness, and nonlinear mapping. In recent years, ANN has been applied and developed to a certain extent in the simulation of river flow and the calculation of 2D plane flow fields. Dibike *et al.* [13] combined the ANN theory and the hydrodynamic model; the hydrodynamic model is used to provide learning samples for the ANN model, and the trained ANN is then adopted to predict the navigation depth, and flow motion in some important areas of the 2D flow field, such as water level, flow velocity, flow direction and flow rate. Yang *et al.* [14] generalized the whole basin into several reservoirs and used the reservoir water balance principle and ANN to simulate the runoff of the Irwell River. The model shows a preferable simulation effect for the daily and monthly runoff time series. The ANN is also adopted to derive operating policies of a reservoir, such as [15–18]. Neelakantan and Pundarikanthan [19] presented a planning model for reservoir operation using a combined backpropagation neural network simulation–optimization (Hooke and Jeeves nonlinear optimization method) process. The combined approach was used for screening the operation policies. Chandramouli and Deka [20] developed a decision support model (DSM) combining a rule based expert system and ANN models to derive operating policies of a reservoir in southern India, and the authors concluded that DSM based on ANN outperforms regression based approaches.

Although some studies of ANN have been published in both hydrodynamics and reservoir operations, respectively, its research and application on reservoir sediment erosion and deposition is still deficient.

In this study, the BP-ANN model is used to determine the complex non-linear relationship between reservoir sediment flushing efficiency and its influencing factors. On the basis of analysis of 1D unsteady flow and sediment mathematical models, four factors composed of reservoir inflow, incoming sediment concentration, reservoir water level, and reservoir release were selected as the input of the model. As the output of the model, sediment flushing efficiency is used to estimate the simulative and predicting accuracy of the model, which should be as close to the desired values as possible. The historical data of the TGR from 2003 to 2010 are used to train the network, and the data in the year of 2011 is adopted for testing. The results indicate that the established model is able to capture the main feature of the reservoir sediment flushing, especially in the flood season, when the majority of the annual sediment is produced. Although the model can be improved with a larger number of samples, the method is proven to be valid and effective.

2. Methodology

The nonlinear mapping of ANN is able to reflect the complex relationship between multiple independent and dependent variables in reservoir sediment flushing with high simulative accuracy and great feasibility. Considering the computational efficiency and practicality, this study adopts ANN to simulate reservoir sedimentation and predict the amount of flushing sediment with different reservoir operational schedules.

2.1. Influence Factors of the Sediment Flushing

Sediment flushing efficiency of reservoir λ shown in Equation (1) is chosen as the indicator of the flushed sediment:

$$\lambda = \frac{S_{out}}{S_{in}},\tag{1}$$

where λ is the reservoir sediment flushing efficiency; S_{out} is the flushed sediment amount out of the reservoir; and S_{in} is the sediment inflow into the reservoir.

Referring to the 1D unsteady flow and sediment mathematical model, the major factors affecting sediment flushing efficiency are selected. The model includes: flow continuity equation as shown in Equation (2), flow motion equation as shown in Equation (3), sediment continuity equation as shown in Equation (4), and the riverbed deformation equation as shown in Equation (5):

$$\frac{\partial Q}{\partial x} + \frac{\partial A}{\partial t} = 0,\tag{2}$$

$$\frac{\partial Q}{\partial t} + \frac{\partial}{\partial t}\left(\frac{Q^2}{A}\right) + gA\frac{\partial H}{\partial x} + g\frac{n^2 Q|Q|}{AR^{\frac{4}{3}}} = 0,\tag{3}$$

$$\frac{\partial AS}{\partial t} + \frac{\partial QS}{\partial x} = -\alpha Bw\left(S - S_*\right),\tag{4}$$

$$\gamma_s\frac{\partial A_s}{\partial t} = \alpha wB\left(S - S_*\right),\tag{5}$$

where

$$H = L - T,\tag{6}$$

$$w = \sqrt{\left(13.95 - \frac{v}{d}\right)^2 + 1.09\frac{\gamma_s}{\gamma}gd} - 1.95\frac{v}{d},\tag{7}$$

$$S_* = \kappa\left(\frac{U^3}{gRw}\right)^l,\tag{8}$$

and γ, γ_s are the volume-weight of water and sediment, respectively; d is the sediment grain size; A is the cross-sectional area; R is the hydraulic radius; Q is the flow; B is the cross-sectional width; H is the water head between upstream and downstream; L is the reservoir water level; T is the tailwater elevation; α is the coefficient of saturation recovery; w is the sediment settling velocity; S is the sediment concentration; S_* is the sediment-carrying capacity of flow; κ, l are coefficient and exponent; and v is the kinematic coefficient of viscosity.

The sediment inflow into a reservoir differs from month to month. A majority of the sediment comes in flood season with flood, and, in non-flood season, the incoming sediment is far less. To improve the simulative accuracy, the sediment delivery characteristics are studied separately for different time periods in a year. Thus, the parameters regarding the water and sediment properties, such as volume-weight, sediment grain size, and kinematic coefficient of viscosity, are considered to be changeless. For each period, the reservoir operation water level is relatively fixed by the operation rules. Thus, the geometric parameters in Equations (2)–(7), including cross-sectional area, width, and hydraulic radius can also be regarded as constants.

On the basis of the analysis above, it can be seen that the major factors that impact the sediment flushing efficiency of reservoirs include: inflow, sediment inflow, release, and water head.

2.2. Artificial Neural Network (ANN)

2.2.1. Outline of ANN

ANN simulates the reaction process of the biological nervous system to information stimulation. To sum up, information processing of neurons consist of two phases: in phase one, the neurons receive information and weight them, known as the integration process; in phase two, the neurons process the integrated information by linear or nonlinear functions, called the activation process. The whole process of inputting information and outputting response can be represented by an activation transfer equation, as shown in Equation (9):

$$Y = F \left(\sum_{i=1}^{n} X_i W_i \right),$$

(9)

where Y is the output of neuron; F is the response characteristic of the neuron to input information; X_i is the input information corresponding to the i-th node; and W_i is the weight of the i-th node.

A neuron is the basic processing unit in neural networks, which is generally a non-linear element with multiple inputs and a single output. Besides being affected by external input signals, the output of a neuron is also influenced by other factors inside the neuron, so an additional input signal θ called bias or threshold is often added in the modeling of artificial neurons. Figure 1 describes the information processing of a single neuron mathematically, where j is the index of the neuron.

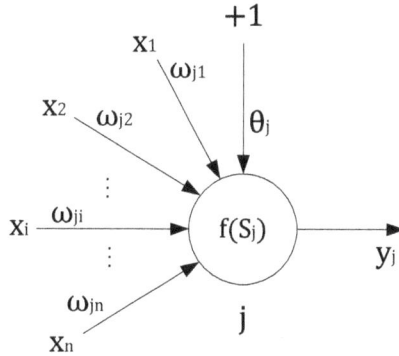

Figure 1. Mathematical model of a single neuron.

A neural network is composed of an input layer, several hidden layers and an output layer, and there are a certain number of neurons in each layer, which are also called nodes.

Neuron mapping between different layers is realized by the activation transfer function. The activation transfer function converts the input in an unlimited domain to output within a limited range. A typical activation transfer function includes: threshold function, linear function, S-type function, hyperbolic tangent function, and so on. The domain divided by an S-type activation transfer function is composed of non-linear hyperplanes, which has a soft, smooth, and arbitrary interface. Thus, it is more precise and rational than the linear function with better robustness. Furthermore, since the S-type function is continuously differentiable, it can be strictly calculated by a gradient method. Equation (10) presents the S-type function:

$$f(x) = \frac{1}{1 + e^{-tx}}. \tag{10}$$

2.2.2. Error Back Propagation Training (BP)-ANN

Neural network algorithms can be classified into the Error Back Propagation Training (BP)-ANN model, the perceptron neural network model, the Radial Basis Function (RBF) neural network model, the Hopfield feedback neural network model, the self-organizing neural network model, and so on. Since it successfully solved the weight adjustment problem of the multilayer feedforward neural network for non-linear continuous functions, the BP-ANN is widely used. Figure 2 shows a mathematical model of the BP-ANN, in which j is the index of the nodes.

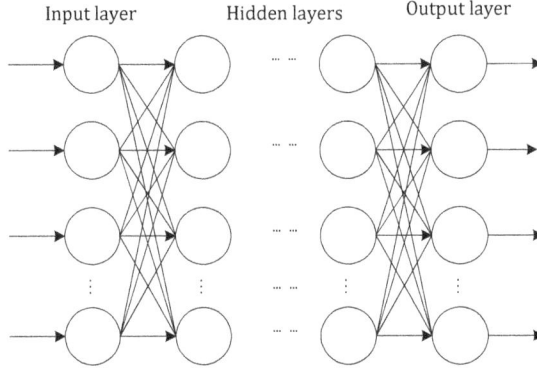

Figure 2. Calculation principle diagram of Back Propagation Training Artificial Neural Network (BP-ANN).

BP-ANN is based on gradient search technology with two processes: forward propagation process of information and back propagation process of errors. In forward propagation, the input signal passes through the hidden layers to the output layer. If desired output appears, the learning algorithm ends; otherwise, it turns to back propagation. In back propagation, the error transits through the output layer, and the weights of each layer of neurons are adjusted according to the gradient descent method to reduce the error. Such loop computation continues to make the output of the network approach the expected values as close as possible. Due to its strong nonlinear mapping ability, BP-ANN is used in this study to establish the relationships between the reservoir flushed sediment and its influence factors, the calculation steps of which are illustrated below.

Assuming the input vector is u, the number of neurons in input layer is n, the output vector is y, the number of neurons in the output layer is m, the length of the input/output sample pair is $\bar{L}$, and the steps of the BP-ANN algorithm include:

(1) Setting the initial weight $\omega\,(0)$, which is relatively small random nonzero value;
(2) Giving the input/output sample pair, and calculating the output of the neural network:

Assuming the input of the p-th sample is $u_p = (u_{1p}, u_{2p}, \ldots, u_{np})$, the output of the p-th sample is $d_p = (d_{1p}, d_{2p}, \ldots, d_{mp})$, $p = 1, 2, \ldots, \bar{L}$, the output of node i with p-th sample is y_{ip}:

$$y_{ip}\,(t) = f\left[x_{ip}\,(t)\right] = f\left[\sum_{j} \omega_{ij}\,(t)\,I_{jp}\right],\tag{11}$$

where I_{jp} is the j-th input of node i when inputting the p-th sample, and $f(\bullet)$ is the activation transfer function as shown in Equation (10);

(3)   Calculating the objective function J of the network:

Assuming E_p is the objective function of the network when inputting the p-th sample, then:

$$E_p = \frac{1}{2} \sum_k \left[d_{kp} - y_{kp}(t) \right]^2 = \frac{1}{2} \sum_k e_{kp}^2(t), \tag{12}$$

where y_{kp} is the network output after t times of weight adjustment when inputting the p-th sample and k is the index of the node in output layer;

The overall objective function of the network is used to estimate the training level of the network, as shown in Equation (13):

$$J(t) = \sum_p E_p(t). \tag{13}$$

(4)   Discriminating whether the algorithm should stop: if $J(t) < \varepsilon$, it stops; otherwise, it turns to step (5), where $\varepsilon > 0$ is preset;
(5)   Back propagation calculating:

Starting from the output layer, referring to J, do the calculation according to the gradient descent algorithm to adjust the value of weight. Assuming the step length is constant, the $(t + 1)$-th adjusted weight of the connection from neuron j to neuron i is:

$$\omega_{ij}(t+1) = \omega_{ij}(t) - \eta \frac{\partial J(t)}{\partial \omega_{ij}(t)} = \omega_{ij}(t) - \eta \sum_p \frac{\partial E_p(t)}{\partial \omega_{ij}(t)} = \omega_{ij}(t) + \Delta \omega_{ij}(t), \tag{14}$$

where η is the step length, also called the learning operator,

$$\frac{\partial E_p}{\partial \omega_{ij}} = \frac{\partial E_p}{\partial x_{ip}} \cdot \frac{\partial x_{ip}}{\partial \omega_{ij}}, \tag{15}$$

$$\delta_{ip} = \frac{\partial E_p}{\partial x_{ip}}, \tag{16}$$

where δ_{ip} is the sensitivity of the status of the i-th node x_{ip} to E_p when inputting the p-th sample. Equation (17) can be derived from Equation (15) to Equation (16):

$$\frac{\partial E_p}{\partial \omega_{ij}} = \delta_{ip} I_{jp}. \tag{17}$$

Calculating δ_{ip} in two different conditions:

① If i is output node, *i.e.*, $i = k$, it can be derived from Equations (12) and (16) that

$$\delta_{ip} = \delta_{kp} = \frac{\partial E_p}{\partial x_{kp}} = \frac{\partial E_p}{\partial y_{kp}} \cdot \frac{\partial y_{kp}}{\partial x_{kp}} = -e_{kp} f'\left(x_{kp}\right). \tag{18}$$

Substitute Equation (18) into Equation (15), then

$$\frac{\partial E_p}{\partial \omega_{ij}} = -e_{kp} f'\left(x_{kp}\right) I_{jp}. \tag{19}$$

② If i is not an output node, *i.e.*, $i \neq k$, Equation (16) is:

$$\delta_{ip} = \frac{\partial E_p}{\partial x_{ip}} = \frac{\partial E_p}{\partial y_{ip}} \cdot \frac{\partial y_{ip}}{\partial x_{ip}} = \frac{\partial E_p}{\partial y_{ip}} \cdot f'\left(x_{ip}\right), \tag{20}$$

in which

$$\frac{\partial E_p}{\partial y_{ip}} = \sum_{m_1} \frac{\partial E_p}{\partial x_{m_1 p}} \cdot \frac{\partial x_{m_1 p}}{\partial y_{ip}} = \sum_{m_1} \frac{\partial E_p}{\partial x_{m_1 p}} \cdot \frac{\partial \sum_j \omega_{m_1 j} I_{jp}}{\partial y_{ip}} = \sum_{m_1} \frac{\partial E_p}{\partial x_{m_1 p}} \cdot \omega_{m_1 i} = \sum_{m_1} \delta_{m_1 p} \cdot \omega_{m_1 i}, \tag{21}$$

where m_1 is the m_1-th node in the layer after node i; I_{jp} is the j-th input for node i. When $i = j$, $y_{jp} = I_{jp}$. Substitute Equations (20) and (21) into Equation (15), then

$$\frac{\partial E_p}{\partial \omega_{ij}} = f\left(x_{ip}\right) I_{jp} \sum_{m_1} \frac{\partial E_p}{\partial x_{m_1 p}} \cdot \omega_{m_1 i} = f\left(x_{ip}\right) I_{jp} \sum_{m_1} \delta_{m_1 p} \cdot \omega_{m_1 i}. \tag{22}$$

Equations (19) and (22) can be used to adjust the weight in Equation (14).

The expression of BP-ANN mapping is a compound of a simple nonlinear function. Several such compound is able to represent complex functions, and then describe many complex processes in physical phenomena.

2.2.3. BP-ANN Model of Reservoir Sedimentation

When combining reservoir sedimentation calculation and reservoir operation, the computational process and constraints are complex, and the time scale is difficult to match. The nonlinear mapping ability of BP-ANN can reflect such complicated relationships between multiple independent and dependent variables without requirements of time-consuming computations.

The sediment flushing efficiency shown in Equation (1) is selected as the output of the BP-ANN model. Based on the analysis in Section 2.2.1, its main influencing factors in real time include: inflow upstream Q_{in}, sediment concentration upstream S, release flow downstream Q_{out}, and reservoir water level L, which are the inputs of the BP-ANN model. Taking the TGR reservoir as an example, a BP-ANN model to simulate and predict the reservoir sediment flushing is established with the observed data from 2003, when the TGP was put into operation, to 2011.

3. Results and Discussion

Hecht-Nielsen [21] proved that for any mapping G from closed unit cube in n-dimensional Euclidean space $[0,1]^n$ to m-dimensional Euclidean space R^m in L_2 (we say a function belongs to L2 if each of its coordinate functions is square-integrable on the unit cube), there exists a three-layer BP neural network, which is capable of approaching G with arbitrary precision. Bourquin $et\ al.$ [22] believed that the Multiple Layer Perception (MLP) with only one hidden layer is sufficient in theory, and, compared to three layers, ANN with four layers is more likely to fall into local optima and harder to train, both of them are similar in other aspects. Hence, BP-ANN model with three layers is selected in this study.

In the application of ANN, it is difficult but crucial to determine an appropriate number of neurons in hidden layer. If the number is too small, the accuracy of the ANN cannot be guaranteed; while if the number is too large, not only the number of connection weights in the network increases, but also the generalization performance of the network is likely to drop. Hence, in this study, the number of neurons in hidden layer is set to be 10.

Input parameters are firstly normalized to fall into [0,1], as shown in Equation (23):

$$\overline{x_i} = \frac{x_i - x_{min,i}}{x_{max,i} - x_{min,i}}. \tag{23}$$

To prevent supersaturation of the neurons, only a certain range r of the S-type activation transfer function with larger curvature is used for mapping, as illustrated in Figure 3.

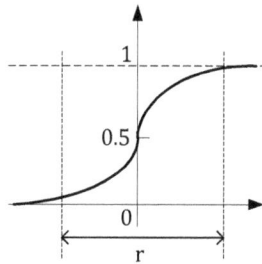

Figure 3. S-type activation transfer function and its adopted range.

Thus, Equation (23) is converted as:

$$\overline{x_i} = \frac{x_i - x_{min,i}\left(1 + d_1\right)}{x_{max,i}\left(1 + d_2\right) - x_{min,i}\left(1 + d_1\right)}. \tag{24}$$

67

After several testing trials of calculation, $d_1 = 40\%$ and $d_2 = 20\%$ in this study.

The four types of observed data $(\vec{Q_{in}}, \vec{Q_{out}}, \vec{L}, \vec{S})$ and the observed sediment flushing efficiency of the TGR from the year 2003 to 2010 are used to train the BP-ANN model, and the data of the year 2011 is used to test it. The simulation and prediction results of the model are shown in Figure 4.

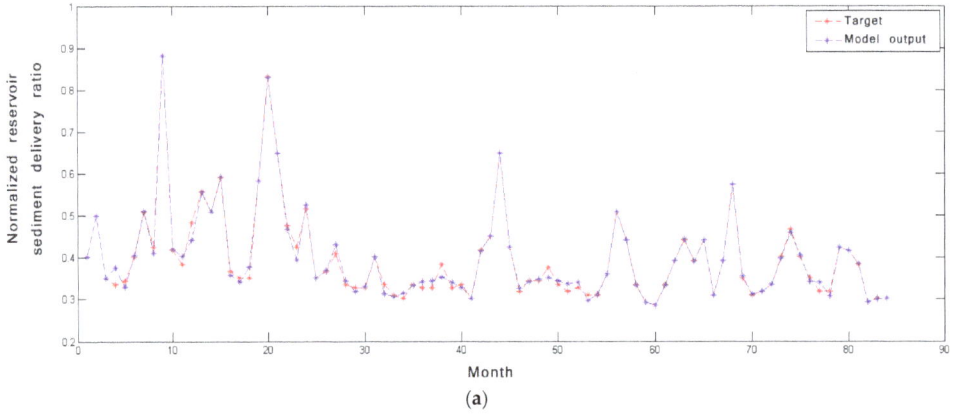

(a)

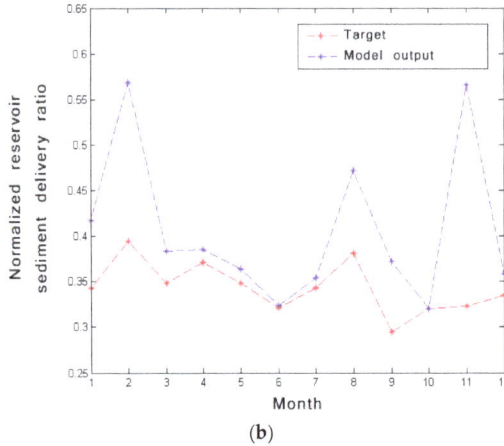

(b)

Figure 4. (a) simulation result and (b) prediction result of the BP-ANN model.

It can be seen from Figure 4 that the trained BP-ANN model can simulate the reservoir sediment flushing process with a high accuracy. When it comes to prediction, the BP-ANN model is able to reflect the relational characteristics of the reservoir sediment flushing. However, since the TGR has only been in operation for about 10 years, and the monthly data is available from 2003 to 2011, the samples used to train the BP-ANN are quite limited, and some deviations still exist. In addition, the sediment delivery from reservoir is an extremely complex process relevant to multiple influencing factors. The reservoir sediment flushing model itself is approximate with many empirical parameters. The four factors selected in this study are necessary and dominated but not sufficient. Last but not least, the majority of the sediment flows into the reservoir in flood season, and very little sediment comes in non-flood season. The difference between the sediment inflow can be several orders of magnitude in different seasons. The observed errors of sediment in the non-flood season itself is not as small as that in flood season. Synthesizing all the reasons above, the BP-ANN model established in this study is believed to be able to characterize the relationship between reservoir flushing sediment and its major influencing factors, especially in flood season, when most of the sediment comes.

Figure 5 shows the goodness of fit between the model output and the observed data. A satisfying fitting degree can be achieved using only the trained data. For the test, the actual data distributes on both sides of the fit line. Basically, the predicted results from the BP-ANN model is a little larger than the actual data, while the gap is limited.

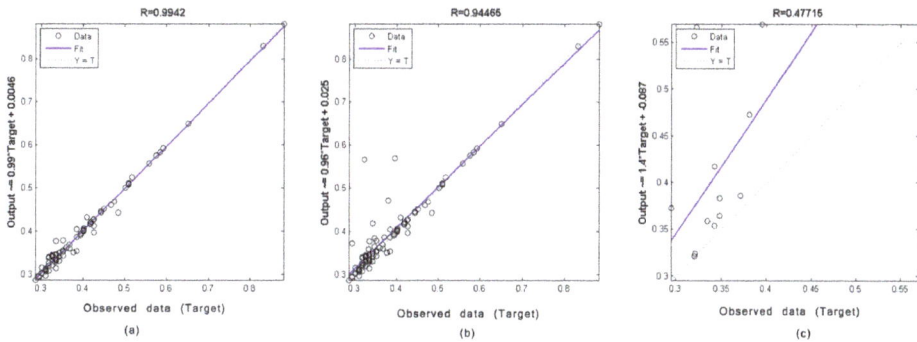

Figure 5. Goodness of fit between the output of the BP-ANN model and the target using (**a**) trained data from 2003 to 2010; (**b**) all the data from 2003 to 2011; and (**c**) tested data of the year 2011.

The statistics of the relative error between the model output and the actual data are shown in Figure 6. There are 96 output variables of the BP-ANN model in the training phase and 12 outputs in testing phase, as the output is monthly sediment flushing efficiency. The blue and green bars in Figure 6 indicate the number of variables with the error falling into a certain range in the training and testing phase, respectively. The red line represents 0 errors occuring. It can be seen that most of the trained outputs stick around the 0 error line. As for the tested outputs, the largest error is about 20%. However, there are only two tested outputs with the error larger than 10%, *i.e.*, over 80% of the predicted sediment flushing efficiency is within 10% difference from the historical data.

As the essential feature can be captured by the BP-ANN model established in this study, it can be inferred that the BP-ANN can be trained with higher prediction accuracy with more samples.

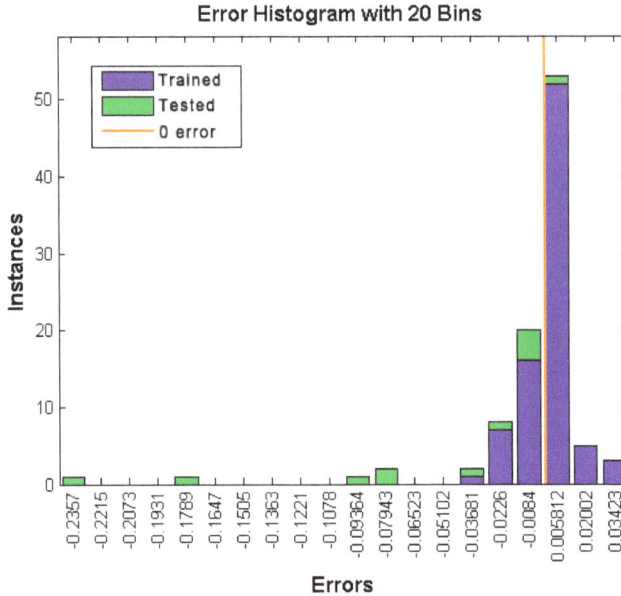

Figure 6. Error histogram with 20 bins of the trained and tested data.

4. Conclusions

Due to the complexity of the reservoir sedimentation process and the immaturity of the sediment research, the calculation of the reservoir sediment erosion and deposition has relied on complicated differential equations with abundant empirical parameters. Since the BP-ANN model has a strong ability to deal with complex non-linear mapping with simpler calculations, it is adopted in order to study the relationships between reservoir flushing sediment and its influencing factors. Four major factors impacting the reservoir sediment flushing efficiency are determined by the analysis on 1D unsteady flow and sediment mathematical model, which are reservoir inflow, water level, outflow, and the inflow sediment concentration. The observed data of the Three Gorges Reservoir from 2003 when it was put into operation to 2010 are used to train the BP-ANN model, and the data of the year 2011 is used for testing. The results show that the established BP-ANN model is able to reflect the essential relationship between the sediment flushing and its influencing factors, especially in flood season, when the majority of the sediment inflows. To improve the accuracy of the model, more observed data is needed in the future.

Acknowledgments: This research was supported by the National Basic Research 973 Program of China (Grant No. 2012CB417002), the National Natural Science Foundation of China (Grant No. 51409248), and the Open Research Fund of State Key Laboratory of Simulation and Regulation of Water Cycle in the River Basin, China Institute of Water Resources and Hydropower Research (Grant No.WHR-SKL-201409).

Author Contributions: Fangfang Li and Jun Qiu conceived and designed the experiments; Xueying Li performed the experiments; Qianqian Shang analyzed the data; Xueying Li and Fangfang Li wrote the paper.

Conflicts of Interest: The authors declare no conflict of interest.

Abbreviations

The following abbreviations are used in this manuscript:

TGP	Three Gorges Reservoir
ANN	Artificial Neural Networks
BP	Back Propagation
MLP	Multiple Layer Perception
1D/2D	1-dimensional/2-dimensional

References

1. Pritchard, S. Overloaded. *Int. Water Power Dam Constr.* **2002**, *54*, 18–22.
2. Palmieri, A.F.; Shah, G.W. Annandale. In *Reservoir Conservation: The RESCON Approach*; World Bank: Washington, DC, USA, 2003.
3. Resources CWRCotMoW. *Chianjiang Sediment Bulletin 2012*; Changjiang Press: Wuhan, China, 2013.
4. Guo, Q.C.; Jin, Y.C. Modeling sediment transport using depth-averaged and moment equations. *J. Hydraul. Eng. ASCE* **1999**, *125*, 1262–1269.
5. Zhou, J.J.; Lin, B.N. One-dimensional mathematical model for suspended sediment by lateral integration. *J. Hydraul. Eng.ASCE* **1998**, *124*, 712–717.
6. Douillet, P.; Ouillon, S.; Cordier, E. A numerical model for fine suspended sediment transport in the southwest lagoon of New Caledonia. *Coral Reefs* **2001**, *20*, 361–372.
7. Aksoy, H.; Kavvas, M.L. A review of hillslope and watershed scale erosion and sediment transport models. *Catena* **2005**, *64*, 247–271.
8. Han, Q.-W. A study of the non-equilibrium transportation of a non-uniform suspended load. *Kexue Tongbao* **1979**, *24*, 804–808.
9. Han, Q.W. Theoretical study of nonequiliirium transportation of nonuniform suspended load. *Water Resour. Hydropower Eng.* **2007**, *38*, 14–23.
10. Liu, W.C.; Lee, C.H.; Wu, C.H.; Kimura, N. Modeling diagnosis of suspended sediment transport in tidal estuarine system. *Environ. Geol.* **2009**, *57*, 1661–1673.
11. Fang, H.W.; Wang, G.Q. Three-dimensional mathematical model of suspended-sediment transport. *J. Hydraul. Eng. ASCE* **2000**, *126*, 578–592.

12. Liu, W.C.; Chan, W.T.; Tsai, D.D.W. Three-dimensional modeling of suspended sediment transport in a subalpine lake. *Environ. Earth Sci.* **2016**, *75*, 173.
13. Dibike, Y.B.; Solomatine, D.; Abbott, M.B. On the encapsulation of numerical-hydraulic models in artificial neural network. *J. Hydraul. Res.* **1999**, *37*, 147–161.
14. Yang, R.F.D.J.; Liu, G.D. Preliminary study on the artificial neural network based on hydrological property. *J. Hydraul. Eng.* **1998**, *8*, 23–27.
15. Raman, H.; Chandramouli, V. Deriving a general operating policy for reservoirs using neural network. *J. Water Resour. Plan. Manag. ASCE* **1996**, *122*, 342–347.
16. Chandramouli, V.; Raman, H. Multireservoir modeling with dynamic programming and neural networks. *J. Water Resour. Plan. Manag. ASCE* **2001**, *127*, 89–98.
17. Cancelliere, A.; Giuliano, G.; Ancarani, A.; Rossi, G. A neural networks approach for deriving irrigation reservoir operating rules. *Water Resour. Manag.* **2002**, *16*, 71–88.
18. Liu, P.; Guo, S.L.; Xiong, L.H.; Li, W.; Zhang, H.G. Deriving reservoir refill operating rules by using the proposed DPNS model. *Water Resour. Manag.* **2006**, *20*, 337–357.
19. Neelakantan, T.R.; Pundarikanthan, N.V. Neural network-based simulation-optimization model for reservoir operation. *J. Water Resour. Plan. Manag. ASCE* **2000**, *126*, 57–64.
20. Chandramouli, V.; Deka, P. Neural network based decision support model for optimal reservoir operation. *Water Resour. Manag.* **2005**, *19*, 447–464.
21. Hecht-Nielsen, R. Theory of the backpropagation neural network. In Proceedings of the International Joint Conference on Neural Networks, IJCNN, Washington, DC, USA, 18–22 June 1989; IEEE: Piscataway, NJ, USA, 1989.
22. Bourquin, J.; Schmidli, H.; van Hoogevest, P.; Leuenberger, H. Basic concepts of artificial neural networks (ANN) modeling in the application to pharmaceutical development. *Pharm. Dev. Technol.* **1997**, *2*, 95–109.

Prediction of the Hot Compressive Deformation Behavior for Superalloy Nimonic 80A by BP-ANN Model

Guo-zheng Quan, Jia Pan and Xuan Wang

Abstract: In order to predict hot deformation behavior of superalloy nimonic 80A, a back-propagational artificial neural network (BP-ANN) and strain-dependent Arrhenius-type model were established based on the experimental data from isothermal compression tests on a Gleeble-3500 thermo-mechanical simulator at temperatures ranging of 1050–1250 °C, strain rates ranging of $0.01–10.0 \text{ s}^{-1}$. A comparison on a BP-ANN model and modified Arrhenius-type constitutive equation has been implemented in terms of statistical parameters, involving mean value of relative (μ), standard deviation (w), correlation coefficient (R) and average absolute relative error ($AARE$). The μ-value and w-value of the improved Arrhenius-type model are 3.0012% and 2.0533%, respectively, while their values of the BP-ANN model are 0.0714% and 0.2564%, respectively. Meanwhile, the R-value and $ARRE$-value for the improved Arrhenius-type model are 0.9899 and 3.06%, while their values for the BP-ANN model are 0.9998 and 1.20%. The results indicate that the BP-ANN model can accurately track the experimental data and show a good generalization capability to predict complex flow behavior. Then, a 3D continuous interaction space for temperature, strain rate, strain and stress was constructed based on the expanded data predicted by a well-trained BP-ANN model. The developed 3D continuous space for hot working parameters articulates the intrinsic relationships of superalloy nimonic 80A.

Reprinted from *Appl. Sci.* Cite as: Quan, G.; Pan, J.; Wang, X. Prediction of the Hot Compressive Deformation Behavior for Superalloy Nimonic 80A by BP-ANN Model. *Appl. Sci.* **2016**, *6*, 66.

1. Introduction

Nimonic 80A, as a nickel-based superalloy, has been widely used in jet engines for aircraft, gas turbines for power plant and marine diesel engines because of its high creep strength, superior oxidation resistance and strong resistance to corrosions at high temperature [1–3]. Generally, the Nimonic 80A is used to fabricate exhausting valve. The upsetting and closed die forging are traditionally applied to form the exhausting valve. However, in recent years, the electric upsetting process with isostatic loading and high heating efficiency is developed to form the exhausting valves [4,5]. It is well known that the hot deformation behavior for a specific material is sensitive to the hot deformation parameters involving strain, strain rate and temperature, and is highly non-linear during hot deformation. The flow behavior of materials is often complex due to the comprehensive function of hardening and softening mechanisms. Consequently, modeling and prediction of the constitutive stress-strain relationships with a high precision is quite complex in nature; meanwhile, it is significant to study and understand the hot deformation behavior and furthermore optimize the deformation process (electric upsetting, forging and extrusion) by numerical simulations. How to obtain an accurate strain-stress relationship becomes critical for the correct calculation of the finite element model [6].

So far, a large amount of research on the characterization for complex non-linear relationships between true stress and deformed parameters such as strain, strain rate and temperature at elevated temperatures has been proposed. Numerous efforts have been made to three types of constitutive models involving the analytical constitutive model, phenomenological constitutive model and artificial neural network [7]. In analytical models, constitutive relations are derived based on physical theories, which require very clear understanding of the processes that control the deformation of the materials. The phenomenological constitutive model is an accurate mathematical model and has relatively many coefficients that need to be calibrated with experimental data. A phenomenological model including the Arrhenius-type equation with hyperbolic laws was proposed to predict flow stress [8]. Furthermore, an improved Arrhenius-type constitutive model incorporating the strain effect on the hot deforming parameters, has been developed to describe and predict the flow behavior for diverse materials or alloys. Lin *et al.* proposed a modified hyperbolic sine constitutive equation, in which the influence of strain was incorporated to predict the flow stress of 42CrMo steel [9]. Later, the modified Arrhenius-type equation was precise for describing the elevated temperature flow stress of Aermet100 steel [10], Ti60 titanium alloy [11], Al–Zn–Mg–Er–Zr alloy [12], *etc.* Such constitutive equations are typically only applicable to the limited materials with specific conditions due to the poor adaptability for the new experimental data. Additionally, the artificial neural network (ANN) model with a back-propagation learning algorithm has been successfully used to predict the hot working behavior of material to overcome the gross approximations introduced by the regression methods [6,13–23]. The back-propagational artificial neural network (BP-ANN) is a model emulating some functions of biological neural networks with a data-driven black-box structure [24], thus it merely needs a collection of some typical examples from the anticipant mapping functions for training regardless of explicit professional knowledge of deformation mechanisms. The BP-ANN model with a data-driven black-box provides a novel way to predict the flow stress by learning the complex and non-linear relationships of flow stress, strain rate, strain and temperature with true stress-strain data. Ji *et al.* applied a feed-forward back-propagation ANN model to predict the flow stress of Aermet100 steel [13]. Haghdadi *et al.* developed a feed-forward back propagation ANN with single hidden layer to predict the flow behavior of an A356 aluminum alloy [15]. Several such works reveal that the predicted results are well consistent with experimental results; furthermore, the neural network is an effective tool to predict the hot deformation behavior of non-linear characteristic materials.

Accordingly, in this work, the stress-strain data of superalloy nimonic 80A were obtained from a series of isothermal compression tests carried out in a wide temperature range of 1050–1250 °C and strain rate range of 0.01–10 s^{-1} on a Gleeble 3500 thermo-mechanical simulator (Dynamic Systems Inc., New York, NY, United States). A BP-ANN model which takes temperature (T), strain rate ($\dot{\varepsilon}$) and strain (ε) as the input variables, and true stress (σ) as the output variable was established by determining proper network structure and parameters to predict the non-linear complex flow behaviors. Meanwhile, a strain-dependent Arrhenius-type constitutive model was constructed to predict the flow stress of nimonic 80A. Subsequently, a comparative analysis on the performance of two such models has been carried out by a series of evaluators such as relative error (δ), average absolute relative error ($AARE$) and correlation coefficient (R), which predictably indicates that the former has higher prediction accuracy. In the following, as described previously, a 3D continuous interaction space within the temperature range of 950–1250 °C, strain rate range of 0.01–10 s^{-1}, and strain range of 0.1–0.9 was constructed.

2. Materials and Experimental Procedure

The chemical compositions (wt. %) of superalloy nimonic 80A used in this study were as follows: C—0.069, Mn—0.630, Cr—2, Fe—1.260, Ti—2.070, Al—0.680, Si—0.550, S—0.001. Twenty nimonic 80A specimens with a diameter of 10 mm and a height of 12 mm were processed from the same extruded billet by wire-electrode cutting. A thermo-mechanical simulator, Gleeble-3500, with a high speed heating system, a servo hydraulic system, a digital control system and a data acquisition system, was used for compression testing. It is common to be used for simulating both mechanical and thermal process at a wide range during hot deformation. Twenty specimens were resistance heated to a proposed deformation temperature with a heating rate of 5 °C/s and then held at that temperature for 180 s by thermo-coupled-feedback-controlled AC current to obtain a homogeneous temperature field. Afterwards, all twenty-four specimens were compressed to a true strain 0.9163 (a fixed height reduction of 60%) at five different temperatures of 1050 °C, 1100 °C, 1150 °C, 1200 °C and 1250 °C, and four different strain rates of 0.01 s^{-1}, 0.1 s^{-1}, 1 s^{-1} and 10 s^{-1} [25]. After each compression, the deformed specimen was immediately quenched into water to retain the high temperature microstructures.

During the compression process, the variations of strain and stress were continuously monitored by the computer equipment with the automatic data acquisition system. Generally, the true stain and true stress were derived from the nominal stress-strain relationship based on the following formula: $\sigma_T = \sigma_N(1 + \varepsilon_N)$, $\varepsilon_T = \ln(1 + \varepsilon_N)$, where σ_T is true stress, σ_N is nominal stress, ε_T is true strain and ε_N is nominal strain.

3. Flow Behavior Characteristics of Superalloy Nimonic 80A

The true compressive stress-strain curves for nimonic 80A, heat-resisting alloy, are illustrated in Figure 1a–d, which show that both deformation temperatures and strain rates have considerable influence on the flow stress of nimonic 80A heat-resisting alloy.

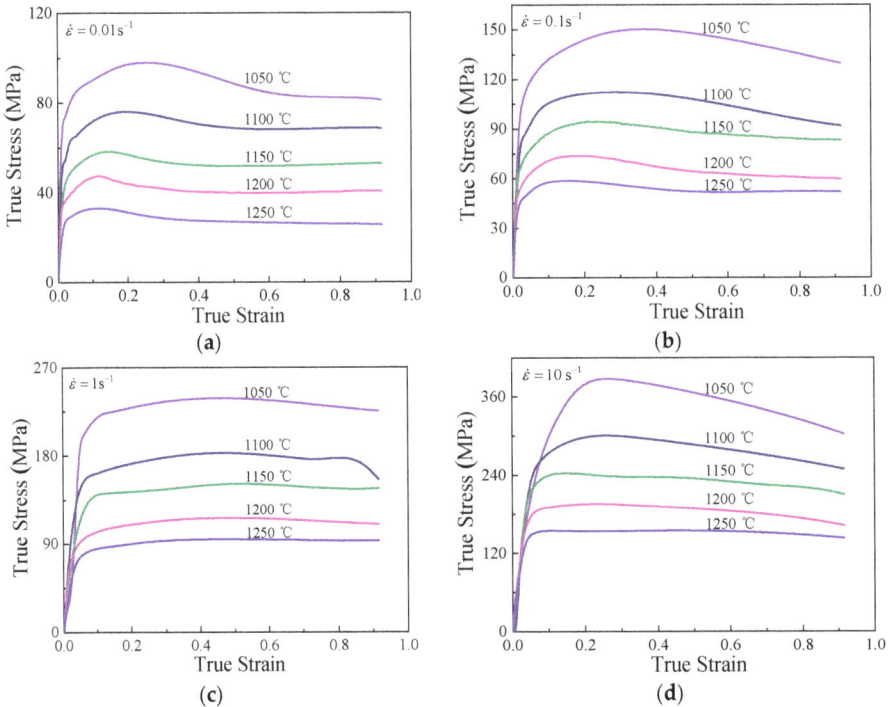

Figure 1. The true stress-strain curves of superalloy nimonic 80A under the different temperatures with strain rates (a) 0.01 s^{-1}; (b) 0.1 s^{-1}; (c) 1 s^{-1}; (d) 10 s^{-1}.

As shown in Figure 1, the strain rate and temperature have a significant effect on the flow curves. Apparently, the flow stress decreases markedly as the temperature increases at a specific strain rate. In contrast, the flow stress increases with the increasing of the strain rate while for a fixed temperature, which is owing to an increase of the dislocation multiplication rate and dislocation density [4]. All the true strain-stress can be summarized in three distinct stages of the stress evolution with strain [4,6,7]. At the first stage of the forming process, the flow stress rapidly increases to a critical value, where work hardening (WH) predominates. At the second stage, where the thermal softening owing to dynamic recrystallization (DRX) and dynamic recovery (DRV) gets more and more predominant, flow stress slowly increases to the peak value even exceeds work hardening. At the third stage, the curves can be divided into two types based on the variation tendency. Evidence of DRX softening, the flow stress decreases continuously, which corresponds to the conditions of 0.01 s^{-1} and 1050–1250 °C, 0.1 s^{-1} and 1050–1200 °C and 1–10 s^{-1} and 1050–1200 °C. However, in the parameter domains of 0.1–10 s^{-1} and 1250 °C, the stress approximately keeps a steady state with significant DRV softening. From the previous descriptions, the typical form of flow curve with DRX softening involved a single peak followed by a flow of steady state. The reason lies in the fact that the highter rate of work hardening slows down the DRX softening rate with lower temperatures and higher strain rates, therefore, the onset of steady state flow is shifted to higher levels [4].

4. Development of Constitutive Relationship for Superalloy Nimonic 80A

4.1. BP-ANN Model

BP-ANN has been widely used to process complex non-linear relationships among several variables [6,20–23,26]. It is a quite efficient computing tool to learn and predict the hot deformation behavior between inputs and outputs by simulating the neural networks structure of the biological neurons. The typical artificial neural network contains three layers, which are input layer, hidden layer and output layer. The input layer receives outside signals and then the output layer generates output signals, while the hidden layer provides the complex network architecture to mimic the non-linear relationship between input signals and output signals [20]. Basically, a feed forward network, which was trained by the back propagation algorithm, was used to establish the back-propagation (BP) neural network. Back-propagation (BP) algorithm adjusts the biases and weights aiming to minimize the target error through gradient descent during training procedure, while learning the relationships between input data and output data.

In this investigation, the input variables of BP-ANN include deformation temperature (T), strain rate ($\dot{\varepsilon}$) and strain (ε), while the output variable is flow stress (σ). The schematic representation of the BP-ANN architecture was shown in Figure 2. All the data from twenty stress-strain curves were divided into training dataset and independent test dataset. In order to ensure the efficiency of training, each continuous stress-strain curve was discretely handled by strain parameter from 0.05 to 0.9 at an interval of 0.01. Hence, a total of 1386 discrete data points from the eighteen stress-strain curves were defined as the training data of this BP-ANN work. The testing dataset was determined as the others curves involving the curves under 0.01 s^{-1} and 1100 °C and 1 s^{-1} and 1200 °C. Among such two curves, the stress values of 36 points picked out from 0.05 to 0.9 with a strain interval of 0.05, and 162 points from the other eighteen training curves in a strain range of 0.05 to 0.9 with a strain interval of 0.1 were considered as the test data for the BP-ANN work performance. The BP-ANN model was trained based on the training dataset, and generalization property of the trained network was assessed by the test dataset selected with a fixed strain rate.

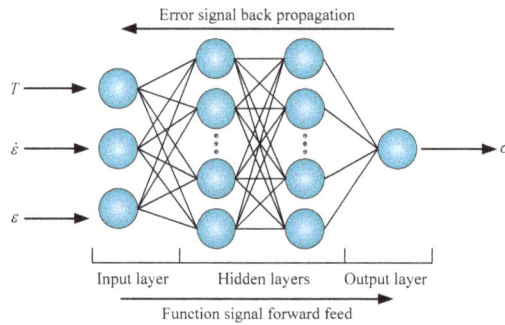

Figure 2. The schematic of the back-propagational artificial neural network architecture.

The selected experimental data have been measured in different units, and thus different data have great differences, which induce the poor convergence speed and predicted accuracy of a BP-ANN model. From the stress-strain curves, it can be seen that the input strain data varies from 0.05 to 0.9, strain rate data varies from 0.01 to 10 s^{-1}, and temperature data varies from 1050 to 1250 °C, the output flow stress data varies from 25.93 MPa to 387.63 MPa. Therefore, before training the network, the input and output datasets have been normalized to avoid value concentrating on weights and some neurons when the iterative calculation of BP-ANN. The main reason for normalizing the data matrix is to recast them into the dimensionless units to remove the arbitrary effect of similarity between the different data. In this research, the normalization processing was realized by Equation (1) [6,17]. The coefficients of 0.05 and 0.25 in Equation (1) are regulating parameters for the sake of narrowing the magnitude of the normalized data within 0.05 to 0.3. Furthermore, it should be noted that the initial numerical values of true stain rates exhibit great magnitude distinction, thereby a logarithm was taken for transforming the true stain rate data before normalization processing.

$$x_n = 0.05 + 0.25 \times \frac{x - 0.95x_{\min}}{1.05x_{\max} - 0.95x_{\min}} \tag{1}$$

where x is the initial data of input or output variables; $x_{\min}$ is the minimum value of x and $x_{\max}$ is the maximum; x_n is the value of x after normalization processing.

The structural parameter settings of the BP-ANN are very complex, which require an appropriate transfer function and an appropriate number of neurons for the hidden layers. It has been proved that two hidden layers are necessary to construct the BP-ANN model to ensure the training accuracy. The determination of the neurons number for hidden layers has a direct relationship with the number of training samples which are often settled by the experience of designers and a trail-and-error procedure. In order to achieve the proposed accuracy, the BP-ANN model was trained with only two neurons for each hidden layer at the beginning; afterwards, the neuron number was adjusted continually (three, four, *etc.*). After repeated trials by changing the neuron number, two hidden layers and 11 neurons in each hidden layer are determined for the final network architecture. Here, "trainbr" function and "learngd" function were empirically chosen as the training function and learning function respectively. In the meantime, the transfer function of the hidden layers was assumed as "tansig" function, whereas the output layer adopted "purelin" function. In addition, an evaluator, sum square error (*SSE*) between experimental and predicted values is introduced into this net to check the ability of the ANN training model. *SSE* is expressed as Equation (2) [6]. Here, the proposed accuracy, *i.e.*, the maximum *SSE*-value is set as 0.0001. The work was accomplished by the neural network toolbox available with MATLAB software (R2013b, MathWorks, Natick, MA, United State, 2013).

$$SSE = \sum_{i=1}^{N} (E_i - P_i)^2 \tag{2}$$

where, E_i is the sample of experimental value; P_i is the sample of predicted value by the BP-ANN model; and N is the number of true stress-strain samples.

Based on the well-trained BP-ANN model, the true stress values under experimental conditions, which include the deformation conditions corresponding to the previous training points and test points, were predicted. Figure 3 exhibits the comparisons between the true stresses predicted by BP-ANN model and the corresponding experimental true stresses for superalloy nimonic 80A. Apparently, the predicted true stress decreases with temperature increasing or strain rate decreasing, which is consistent with experimental stress-strain curves. The phenomenon predictably indicates that the BP-ANN model is able to effectively grasp the stress-strain evolution rules, that is, it possesses excellent capability to track the dynamic softening (including DRX and DRV) and work hardening regions of superalloy nimonic 80A. Additionally, the test data including the data under 0.01 s^{-1} and 1100 °C and 1 s^{-1} and 1200 °C, are used to assess the generalization property of the BP-ANN model. The result of comparisons shows that the true stresses predicted by BP-ANN model has good agreement with experimental stress-strain curves, which indicates the high generalization property of the BP-ANN model.

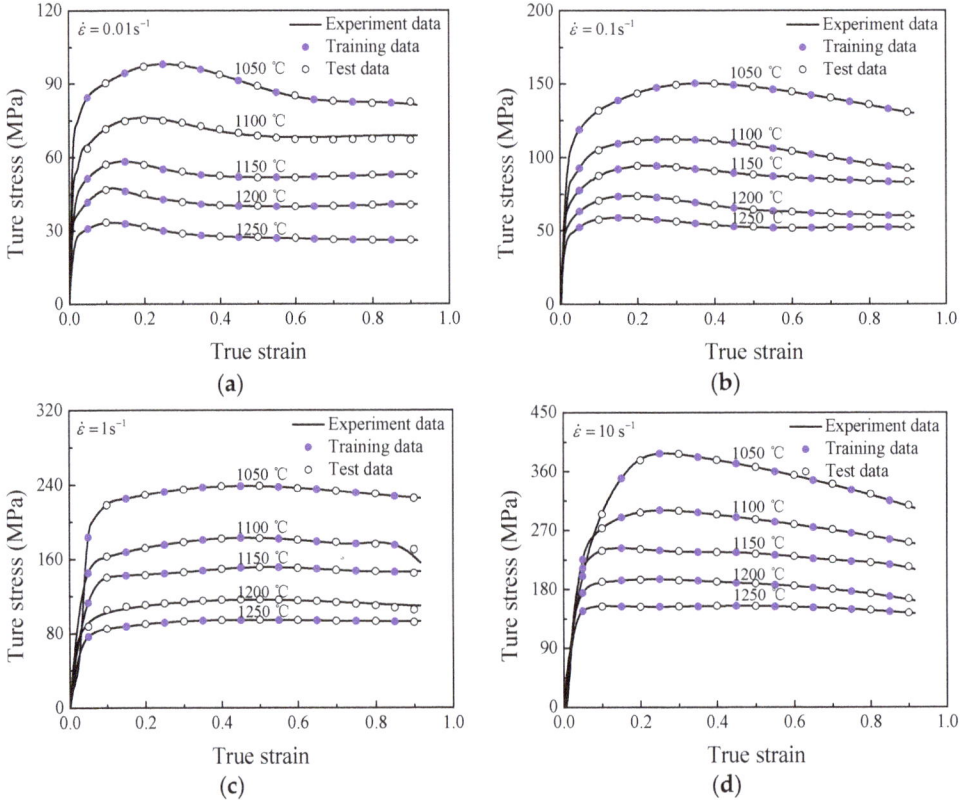

Figure 3. The comparison of the BP-ANN prediction with experimental values at different temperatures and strain rates (a) 0.01 s^{-1}; (b) 0.1 s^{-1}; (c) 1 s^{-1}; (d) 10 s^{-1}.

4.2. Arrhenius-Type Constitutive Model

Generally, Arrhenius type equation is expressed as Equation (3) [27], which correlates the flow stress (σ) with temperature (T) and strain rate ($\dot{\varepsilon}$).

$$\dot{\varepsilon} = AF(\sigma)\exp(-Q/RT) \tag{3}$$

where $F(\sigma) = \begin{cases} \sigma^{n'} & \alpha\sigma < 0.8 \\ \exp(\beta\sigma) & \alpha\sigma > 1.2 \\ [\sinh(\alpha\sigma)]^n & for\ all\ \sigma \end{cases}$,

and where $\dot{\varepsilon}$ is the strain rate (s^{-1}), R is the universal gas constant $(8.31\,J\cdot mol^{-1}\cdot K^{-1})$, T is the absolute temperature (K), Q is the activation energy of deformation $(kJ\cdot mol^{-1})$, σ is the flow stress (MPa) for a given stain, A, α, n' and n are the material constants, $\alpha = \beta/n'$.

For the low stress level $(\alpha\sigma < 0.8)$, taking natural logarithms on both sides of Equation (3), the following equation can be obtained:

$$\ln\dot{\varepsilon} = \ln A + n'\ln\sigma - Q/RT \tag{4}$$

For the high stress level $(\alpha\sigma > 1.2)$, taking natural logarithms on both sides of Equation (3) gives:

$$\ln\dot{\varepsilon} = \ln A + \beta\sigma - Q/RT \tag{5}$$

According to Equations (4) and (5), $n' = d\ln\dot{\varepsilon}/d\ln\sigma$ and $\beta = d\ln\dot{\varepsilon}/d\sigma$. Then, the linear relationships of $\ln\sigma$–$\ln\dot{\varepsilon}$ and σ–$\ln\dot{\varepsilon}$ for strain of 0.5 at the temperatures of 1050–1250 °C were fitted out as shown in Figure 4. The adjusted coefficient of determination R^2 for each condition was calculated and showed in Figure 4, which was used to prove the reliability of fitting curves. The inverse of the slopes of straight lines in $\ln\sigma$–$\ln\dot{\varepsilon}$ and σ–$\ln\dot{\varepsilon}$ plots is accepted as the values of material constants n' and β at each tested temperature, respectively. Thus the values of n' and β at strain of 0.5 were obtained by averaging the inverse of slopes under different temperatures, which were found to be 5.3625 MPa^{-1} and 0.0336 MPa^{-1}, respectively. Furthermore, the value of another material constant $\alpha = \beta/n' = 0.0063$ MPa^{-1} was also obtained.

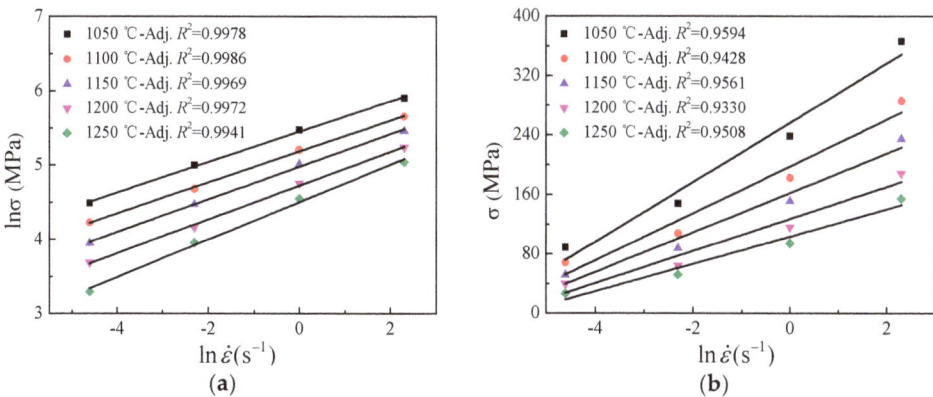

Figure 4. Relationship between (a) $\ln\sigma$ and $\ln\dot{\varepsilon}$ and (b) σ and $\ln\dot{\varepsilon}$.

For all the stress level (including low and high-stress levels), Equation (3) can be rewritten as the following:

$$\dot{\varepsilon} = A[\sinh(a\sigma)]^n \exp(-Q/RT) \tag{6}$$

Taking natural logarithms on both sides of Equation (6), the following equation can be obtained:

$$\ln\dot{\varepsilon} = \{\ln A + n[\ln \sinh(\alpha\sigma)] - Q\}/RT \tag{7}$$

For the given deformation temperature (T), the stress exponent (n) is expressed as Equation (8):

$$n = \frac{\partial \ln\dot{\varepsilon}}{\partial \ln(\sinh(\alpha\sigma))}\Big|_T \tag{8}$$

When the strain rate ($\dot{\varepsilon}$) is a constant, the activation energy (Q) can be expressed as Equation (9):

$$Q = nR\frac{\partial \ln(\sinh(\alpha\sigma))}{\partial(1/T)}\Big|_{\dot{\varepsilon}} \tag{9}$$

According to Equations (8) and (9), the linear relationship between $\ln(\sinh(\alpha\sigma))$ and $\ln\dot{\varepsilon}$ and the relationship between $\ln(\sinh(\alpha\sigma))$ and $1/T$ were fitted out as shown in Figure 5 and the determination coefficient R^2 have been exhibited in each figure. Consequently, the value of constant parameter n and the activation energy Q can be derived from the mean slope of lines in Figure 5a,b respectively, here, n is 3.7806 and Q is 403.81 kJ·mol^{-1}. In addition, the material constant A can be calculated as 4.5496×10^{14} s^{-1}.

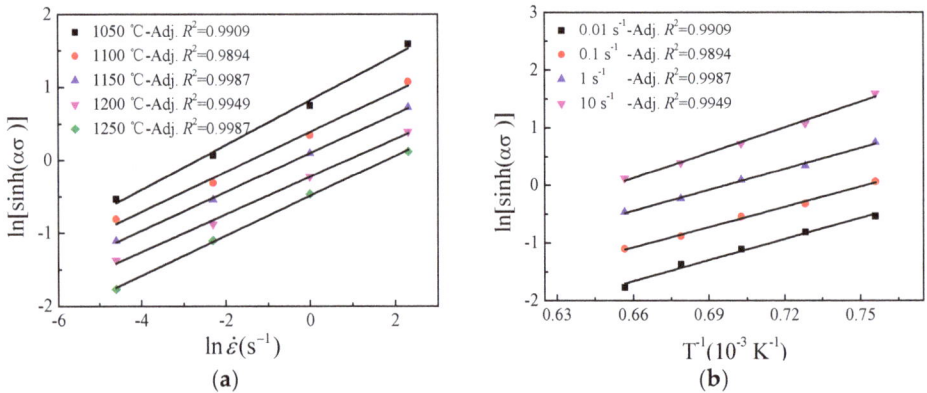

Figure 5. Relationships between: (a) $\ln(\sinh(\alpha\sigma))$ and $\ln\dot{\varepsilon}$ (b) $\ln(\sinh(\alpha\sigma))$ and $1/T$

However, the effect of temperature and strain rate on flow behavior cannot be considered in Equation (3). Zener-Hollomon parameter, Z, in an exponent-type Equation (10) [28] has been introduced to model the comprehensive function of temperature and strain rate.

$$Z = \dot{\varepsilon}\exp(\frac{Q}{RT}) \tag{10}$$

Base on Equations (6) and (10), the stress σ at the strain of 0.5 can be written as a function of Z parameter:

$$\sigma = \frac{1}{\alpha}\ln\left\{ (Z/A)^{\frac{1}{n}} + \left[(Z/A)^{\frac{2}{n}} + 1 \right]^{\frac{1}{2}} \right\} \tag{11}$$

It is well known that the constitutive model is affected not only by the deformation temperature and strain rate, but also by the strain in the hot deformation of metal materials [7,10,12,17,29–32]. Therefore, the values of material coefficients (i.e., α, A, n, Q) of the constitutive equation are calculated under different strains in a range of 0.05 to 0.9 with the interval of 0.05 by the same method used previously. These values were then used to fit the polynomial functions (Figure 6), and the variation of α, $\ln A$, n and Q with true strain ε could be represented by a sixth order polynomial respectively, as shown in Equation (12). The coefficients of the sixth order polynomial functions are tabulated in Table 1:

$$\begin{cases} \alpha = B_0 + B_1\varepsilon + B_2\varepsilon^2 + B_3\varepsilon^3 + B_4\varepsilon^4 + B_5\varepsilon^5 + B_6\varepsilon^6 \\ \ln A = C_0 + C_1\varepsilon + C_2\varepsilon^2 + C_3\varepsilon^3 + C_4\varepsilon^4 + C_5\varepsilon^5 + C_6\varepsilon^6 \\ n = D_0 + D_1\varepsilon + D_2\varepsilon^2 + D_3\varepsilon^3 + D_4\varepsilon^4 + D_5\varepsilon^5 + D_6\varepsilon^6 \\ Q = E_0 + E_1\varepsilon + E_2\varepsilon^2 + E_3\varepsilon^3 + E_4\varepsilon^4 + E_5\varepsilon^5 + E_6\varepsilon^6 \end{cases} \tag{12}$$

Table 1. Polynomial fitting results of superalloy nimonic 80A.

	α		$\ln A$		n		Q
B_0	0.01006	C_0	40.29126	D_0	7.23447	E_0	467.28232
B_1	−0.05783	C_1	−144.12464	D_1	−44.03632	E_1	−1502.21622
B_2	0.33009	C_2	982.95072	D_2	234.75515	E_2	10438.62694
B_3	−0.95142	C_3	−3062.36904	D_3	−653.05658	E_3	−32768.83516
B_4	1.47519	C_4	4851.02761	D_4	984.16398	E_4	52125.14469
B_5	−1.16554	C_5	−3813.23640	D_5	−756.88970	E_5	−41.091.74522
B_6	0.36783	C_6	1183.77777	D_6	232.91924	E_6	12784.92195

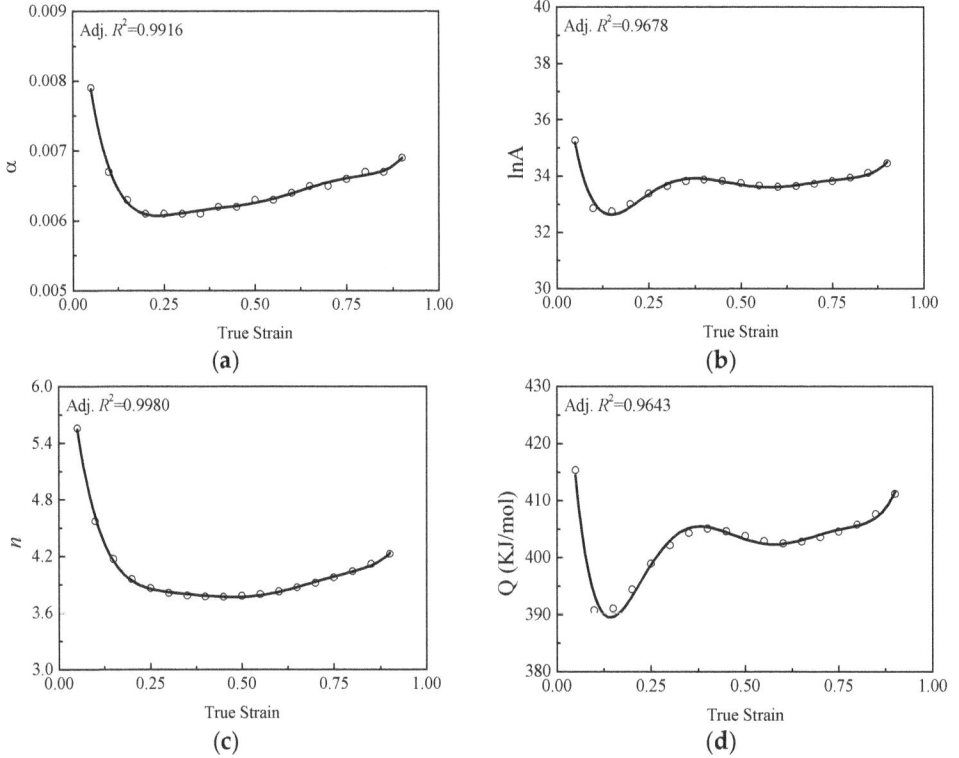

Figure 6. Relationships between: (a) α; (b) $\ln A$; (c) n; and (d) Q and true strain ε by polynomial fit.

Thus, the improved Arrhenius type model with variable coefficients can be expressed as Equation (13).

$$\sigma = \frac{1}{g\left(\varepsilon\right)}\ln\left\{\left(\frac{\dot{\varepsilon}\exp[j(\varepsilon)/8.314T]}{f(\varepsilon)}\right)^{1/h(\varepsilon)} + \left[\left(\frac{\dot{\varepsilon}\exp(j(\varepsilon)/8.314T)}{f(\varepsilon)}\right)^{2/h(\varepsilon)} + 1\right]^{1/2}\right\} \quad (13)$$

where $g\left(\varepsilon\right)$, $f\left(\varepsilon\right)$, $h\left(\varepsilon\right)$, $j\left(\varepsilon\right)$ are polynomial functions of strain for α, A, n, Q.

Applying the aforementioned material constants to Equation (13), the true stress values are calculated for the experimental temperature, strain and strain rate ranges. Figure 7 shows comparisons between the experimental data and the predicted results calculated from the developed constitutive equations (considering the compensation of strain) at the temperatures of 1050 °C, 1100 °C, 1150 °C, 1200 °C, and 1250 °C, and the strain rates of 0.01 s^{-1}, 0.1 s^{-1}, 1 s^{-1} and 10 s^{-1}. It can be seen that the proposed constitutive equation gives an accurate estimation on the flow stress of superalloy nimonic 80A in most of the experimental conditions.

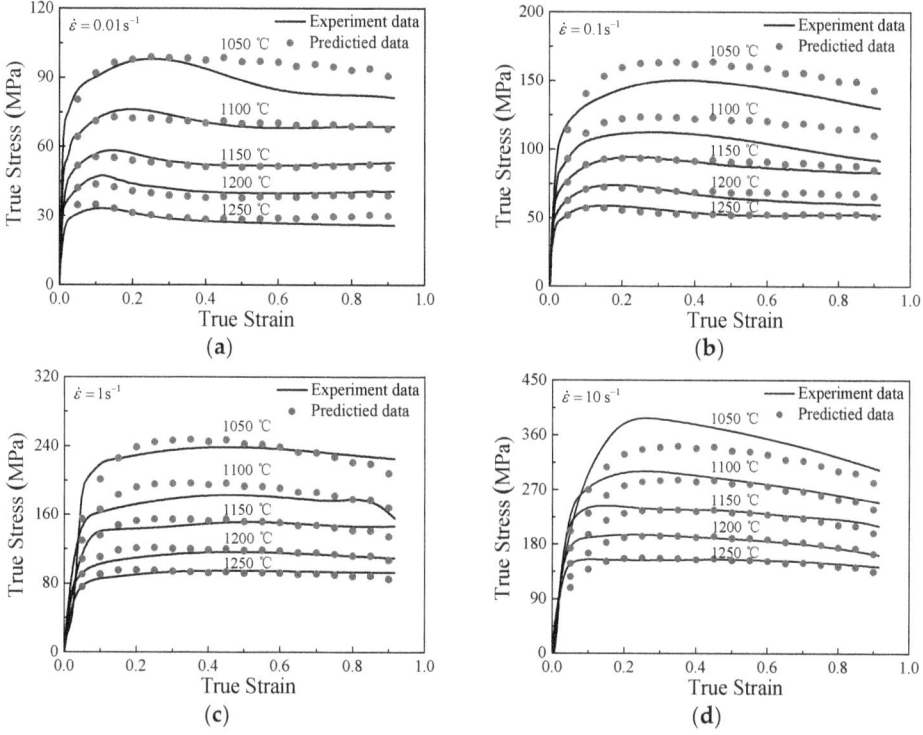

Figure 7. Comparisons between predicted and measured under different deformation temperatures with strain rates of (**a**) 0.01 s^{-1}; (**b**) 0.1 s^{-1}; (**c**) 1 s^{-1} and (**d**) 10 s^{-1}.

5. Prediction Capability Comparison between the BP-ANN Model and Arrhenius Type Constitutive Equation

Depending on the improved Arrhenius type constitutive equation, in this study, the true stresses of 32 points under the conditions of 1100 °C and 0.01 s^{-1} and 1200 °C and 1 s^{-1} at a strain range of 0.05 to 0.9 with a strain interval of 0.05 were calculated to compare with the true stress predicted by BP-ANN model and obtained from the isothermal compression tests. For the sake of the contrast of prediction accuracy between these two models, the relative error (δ) is introduced, which is expressed by Equation (14).

$$\delta\,(\%) = \frac{P_i - E_i}{E_i} \times 100\% \qquad (14)$$

where E_i is the sample of experimental value and P_i is the sample of predicted value.

The δ-values relative to the experimental true stress was calculated by Equation (14) and listed in Table 2.

Table 2. Relative errors of the predicted results by the back-propagational artificial neural network (BP-ANN) model and constitutive equation to experimental results under the condition of 1100 °C and 0.01 s^{-1} and 1200 °C and 1 s^{-1}.

Strain Rate (s^{-1})	Temperature (°C)	Strain	True Stress (MPa)		Equation	Relative Error (%)	
			Experimental	BP-ANN		BP-ANN	Equation
0.01	1100	0.05	65.08	63.59	64.37	−2.29	−1.08
		0.10	71.80	71.58	71.18	−0.31	−0.86
		0.15	75.24	74.75	72.79	−0.64	−3.25
		0.20	76.09	75.31	72.42	−1.03	−4.83
		0.25	75.35	74.99	72.30	−0.47	−4.04
		0.30	73.81	74.05	71.59	0.33	−3.01
		0.35	72.06	72.75	71.23	0.96	−1.15
		0.40	70.57	71.31	70.43	1.04	−0.20
		0.45	69.47	69.93	71.33	0.67	2.68
		0.50	68.73	68.77	70.15	0.05	2.06
		0.55	68.32	67.90	70.40	−0.62	3.04
		0.60	68.17	67.36	70.40	−1.18	3.28
		0.65	68.20	67.13	69.34	−1.56	1.68
		0.70	68.34	67.13	70.24	−1.78	2.77
		0.75	68.54	67.26	69.70	−1.87	1.69
		0.80	68.72	67.39	68.91	−1.93	0.28
		0.85	68.80	67.37	69.76	−2.08	1.39
		0.90	68.73	67.08	67.71	−2.41	−1.49
1	1200	0.05	91.67	87.76	90.72	−4.26	−1.04
		0.10	102.36	105.69	111.12	3.24	8.55
		0.15	106.88	108.91	118.77	1.90	11.13
		0.20	109.66	110.69	121.14	0.94	10.47
		0.25	111.91	112.48	121.60	0.51	8.66
		0.30	113.69	114.08	120.83	0.35	6.28
		0.35	114.99	115.35	120.39	0.31	4.69
		0.40	115.82	116.21	119.02	0.34	2.76
		0.45	116.23	116.63	120.40	0.34	3.59
		0.50	116.27	116.61	118.44	0.30	1.87
		0.55	115.99	116.25	118.69	0.22	2.33
		0.60	115.45	115.62	118.07	0.14	2.26
		0.65	114.71	114.75	115.69	0.04	0.86
		0.70	113.82	113.58	116.16	−0.20	2.06
		0.75	112.83	112.03	114.25	−0.71	1.26
		0.80	111.81	110.07	111.94	−1.55	0.12
		0.85	110.80	107.84	111.96	−2.67	1.05
		0.90	109.86	105.50	107.05	−3.97	−2.56

It is found in Table 2 that the relative percentage error obtained from BP-ANN model varies from −4.26% to 3.24%, whereas it is in the range from −4.83% to 11.13% for the improved Arrhenius-type constitutive model. As shown in Figure 8a,b, the relative percentage errors have been summarized in which the height of the histogram expresses the relative frequency of the relative percentage errors. Through nonlinear curve fitting, the distributions of relative percentage errors obtained from Arrhenius-type model and BP-ANN model present a typical Gaussian distribution, which was expressed as in Equation (15). In the function, the two parameters of μ and w represent the mean value and standard deviation, respectively, which are two of the most important indexes in statistical work. The mean value and standard deviation calculated by Equations (16) and (17) [6], respectively, reflect the central tendency and discrete degree of a set of data, and smaller value of w and μ close to 0 hint that better errors distribution is achieved. As shown in Figure 8a,b, the mean value (μ) and the standard deviation (w) of the Arrhenius-type model is 3.0012 and 2.0533, respectively, while the mean value (μ) and the standard deviation (w) of the BP-ANN model is 0.0714 and 0.2564, respectively, which indicate that the distribution of relative percentage errors obtained by the BP-ANN model is more centralized. It suggests that the BP-ANN model has a good generalization capability.

$$y = y_0 + A e^{-\dfrac{(\delta_i - \mu)^2}{2w^2}} \tag{15}$$

$$\mu = \frac{1}{N} \sum_{i=1}^{N} \delta_i \tag{16}$$

$$w = \sqrt{\frac{1}{N-1} \sum_{i=1}^{N} (\delta_i - \mu)^2} \tag{17}$$

where δ_i is a value of the relative error; μ, w, and y are the mean value, standard deviation and probability density of δ respectively; y_0 and A are constants, and N is the number of relative errors, here $N = 36$.

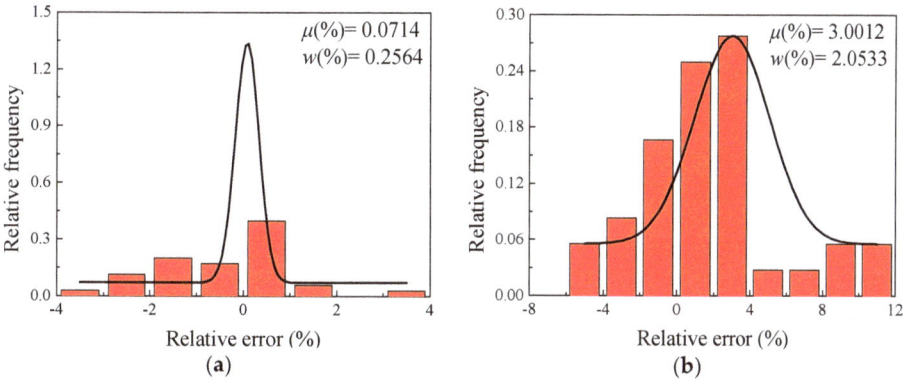

Figure 8. The relative errors distribution on the true stress points predicted by (a) the BP-ANN model and (b) the Arrhenius type constitutive equation relative to the experimental ones.

In addition, two commonly used statistical indicators of the correlation coefficient (R) and average absolute relative error ($AARE$) are introduced to check the ability and the predictability of the BP-ANN model and Arrhenius-type model, which are expressed by Equations (18) and (19) [6]. R is a numerical value between -1 and 1 that expresses the strength of the linear relationship between two variables. A high R-value close to 1 illustrates that the predicted values conform to the experimental ones well. A value of 0 indicates that there is no relationship. Value close to -1 signal a strong negative relationship between the two variables. The average absolute relative error ($AARE$) is also computed through a term-by-term comparison of the relative error and thus is an unbiased statistical parameter to measure the predictability of a model [6,15]. Meanwhile, a low $AARE$-value close to 0 indicates that the sum of the errors between the predicted and experimental values tends to be 0.

$$R = \frac{\sum_{i=1}^{N} (E_i - \overline{E})(P_i - \overline{P})}{\sqrt{\sum_{i=1}^{N} (E_i - \overline{E})^2 \sum_{i=1}^{N} (P_i - \overline{P})^2}} \tag{18}$$

$$AARE\,(\%) = \frac{1}{N} \sum_{i=1}^{N} \left| \frac{P_i - E_i}{E_i} \right| \times 100\% \tag{19}$$

where E_i is the sample of experimental value and P_i is the sample of predicted value. $\overline{E}$ is the mean value of experimental sample values, $\overline{P}$ is the mean value of predicted sample values, and N is the number of data which were employed in the investigation.

The correlation relationships between the experimental and respectively predicted true stress by BP-ANN model and modified Arrhenius-type constitutive equation were illustrated in Figure 9. It is discovered that the points in Figure 9, which take experimental true stress as horizontal axis and predicted true stress as vertical axis, lie fairly close to the best linear fitted line, suggesting that the predicted stress-strain values conform very well to the homologous experimental ones. Besides, the R-values for the predicted true stress of BP-ANN and modified Arrhenius are 0.9998 and 0.9899, respectively, from another quantitative perspective proving the strong linear relationships between the predicted and experimental true stress. Additionally, the $AARE$-values relative to the experimental true stress was calculated by Equation (19) and exhibited in Figure 9. According to the calculation results, it is manifest that the $AARE$-value for the BP-ANN model is 1.20%, but, for the constitutive equation, it reaches a higher level, 3.06%. Lower $AARE$-value means a smaller deviation on the whole; therefore, the BP-ANN model has higher accuracy in predicting the true stress of superalloy nimonic 80A than the constitutive equation.

By several comparison methods, the performance of the two models can be concluded that the BP-ANN model has higher prediction accuracy than the improved Arrhenius-type model. It is valuable to note that, in the training stage of the BP-ANN model, the experimental stress-strain data of two test curves under the conditions of 1100 °C and 0.01 s^{-1} and 1200 °C and 1 s^{-1} did not participate. However, when establishing the constitutive equation, they were involved. However, even on this premise, the BP-ANN model still shows smaller errors, giving the full proof that the present BP-ANN model has better prediction capability than the constitutive equation in the flow characteristics of superalloy nimonic 80A.

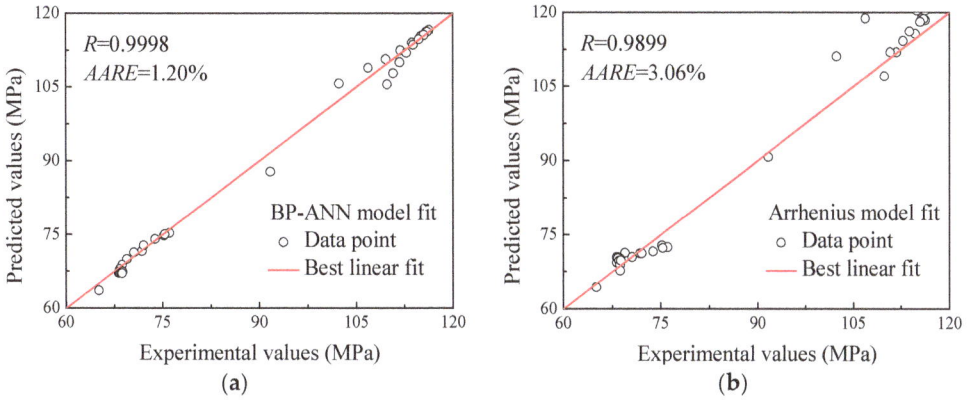

Figure 9. The correlation relationships between the predicted and experimental true stress for the (**a**) BP-ANN model and (**b**) Arrhenius-type model.

6. Prediction Potentiality of BP-ANN Model

There is no doubt that the well-trained BP-ANN model is effective to predict the flow stress based on the experimental data for the non-linear material. With the well-trained BP-ANN model, the flow stresses outside of the experimental conditions including 950 °C and 1000 °C were predicted for superalloy nimonic 80A. Additionally, based on the data at a temperature range of 950 to 1250 °C under strain rate of $0.01\ s^{-1}$, $0.1\ s^{-1}$, $1\ s^{-1}$ and $10\ s^{-1}$, an interpolation method was implemented to densely insert stress-strain data into these data; furthermore, a 3D continuous response space (illustrated in Figure 10) with flow stress along the V-axis and deformation temperature, logarithm of strain rate and strain and along the X, Y and Z axes, respectively, was constructed by a surface fitting process. The values of V-axis are represented by different colors. Figure 10a shows the 3D continuous interaction space, which reveals the continuous response relationship between stress and strain, strain rate and temperature of superalloy nimonic 80A. Figure 10b–d respectively exhibit the cutting slices of 3D continuous response mapping at diverse parameters, involving temperature, strain rate and strain. In the 3D continuous interaction space, all the stress-strain points are digital and can be determined, since the surface fitting step has transformed the discrete stress-strain points into continuous stress-strain surface and space. The accuracy of such a 3D continuous interaction space is strongly guaranteed by the excellent prediction performance of an optimally-constructed and well-trained BP-ANN model. As is known, the stress-strain data are the most fundamental data to predict the deformation behaviors of the superalloy nimonic 80A during electric upsetting with finite element model It is realizable to pick out dense stress-strain data from the 3D continuous interaction space and insert such continuous mapping relationships into commercial software such as Marc, *etc.* by program codes. In this way, the accurate simulation of one certain forming process is able to perform.

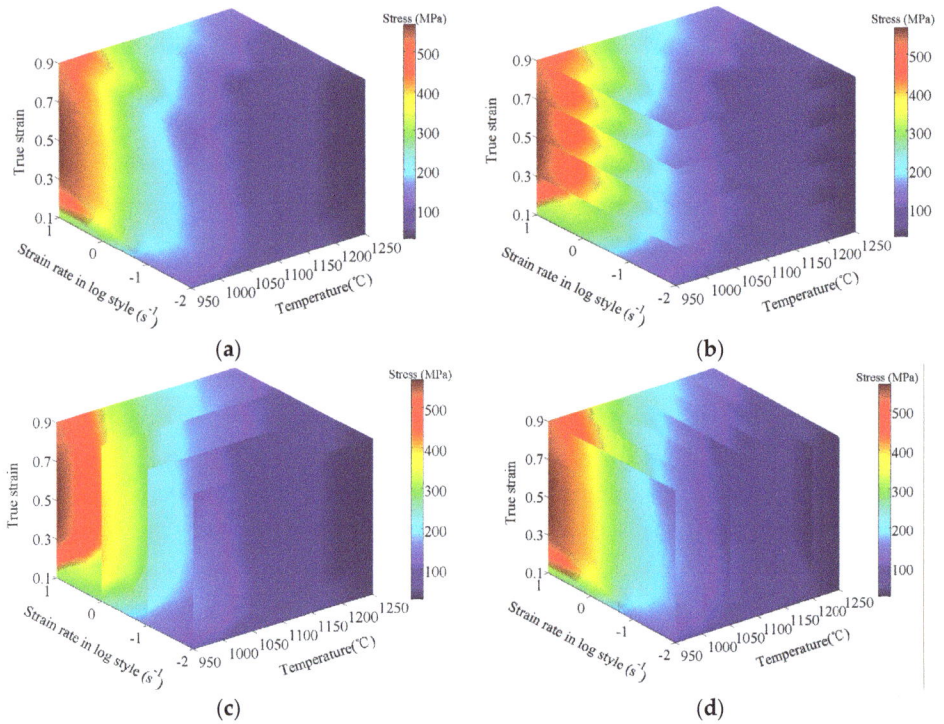

Figure 10. The 3D relationships among temperature, strain rate, strain and stress: (**a**) 3D continuous interaction space, 3D continuous mapping relationships under different (**b**) temperatures; (**c**) strain rates and (**d**) strains.

7. Conclusions

(1) A BP-ANN model taking the deformation temperature (T), strain rate ($\dot{\varepsilon}$) and strain (ε) as input variables and the true stress (σ) as output variable was constructed for the compression flow behaviors of superalloy, nimonic 80A, which presents desired precision and reliability.

(2) A strain-dependent Arrhenius-type model is developed to predict the flow behavior of superalloy nimonic 80A under the specific deformation conditions. A sixth order polynomial is adopted to reveal the relationships between

variable coefficients (including activation energy Q, material constants n, α, and A) and strain with good correlations.

(3) A series of statistical indexes, involving the relative error (δ), mean value (μ), standard deviation (w), correlation coefficient (R) and average absolute relative error ($ARRE$), were introduced to contrast the prediction accuracy between the improved Arrhenius type constitutive equation and BP-ANN model. The mean value (μ) and standard deviation (w) of the improved Arrhenius-type model are 3.0012% and 2.0533%, respectively, while their values of the BP-ANN model are 0.0714% and 0.2564%, respectively. Meanwhile, the correlation coefficient (R) and average absolute relative error ($ARRE$) for the improved Arrhenius-type model are 0.9899 and 3.06%, while their values for the BP-ANN model are 0.9998 and 1.20%, which indicate that the BP-ANN model has a good generalization capability.

(4) The true stress data within the temperature range of 950–1250 °C, the strain rate range of 0.01–10 s^{-1}, and the strain range of 0.1–0.9 were predicted densely. According to these abundant data, a 3D continuous interaction space was constructed by interpolation and surface fitting methods. It significantly contributes to all the research requesting abundant and accurate stress-strain data of superalloy nimonic 80A.

Acknowledgments: This work was supported by the National Natural Science Foundation of China (51305469). The corresponding author was appreciated from the Chongqing Higher School Youth-Backbone Teacher Support Program.

Author Contributions: Guo-zheng Quan and Jia Pan conceived and designed the experiments; Xuan Wang performed the experiments; Jia Pan and Xuan Wang analyzed the data; Xuan Wang contributed reagents/materials/analysis tools; Guo-zheng Quan and Jia Pan wrote the paper.

Conflicts of Interest: The authors declare no conflict of interest.

References

1. Kim, D.K.; Kim, D.Y.; Ryu, S.H.; Kim, D.J. Application of Nimonic 80A to the hot forging of an exhaust valve head. *J. Mater. Process. Technol.* **2001**, *113*, 148–152.
2. Zhu, Y.; Zhimin, Y.; Jiangpin, X. Microstructural mapping in closed die forging process of superalloy Nimonic 80A valve head. *J. Alloys Compd.* **2011**, *509*, 6106–6112.
3. Tian, B.; Zickler, G.A.; Lind, C.; Paris, O. Local microstructure and its influence on precipitation behavior in hot deformed Nimonic 80A. *Acta Mater.* **2003**, *51*, 4149–4160.
4. Quan, G.-Z.; Mao, A.; Luo, G.-C.; Liang, J.-T.; Wu, D.-S.; Zhou, J. Constitutive modeling for the dynamic recrystallization kinetics of as-extruded 3Cr20Ni10W2 heat-resistant alloy based on stress-strain data. *Mater. Des.* **2013**, *52*, 98–107.

5. Quan, G.-Z.; Liang, J.-T.; Liu, Y.-Y.; Luo, G.-C.; Shi, Y.; Zhou, J. Identification of optimal deforming parameters from a large range of strain, strain rate and temperature for 3Cr20Ni10W2 heat-resistant alloy. *Mater. Des.* **2013**, *52*, 593–601.

6. Quan, G.-Z.; Lv, W.-Q.; Mao, Y.-P.; Zhang, Y.-W.; Zhou, J. Prediction of flow stress in a wide temperature range involving phase transformation for as-cast Ti–6Al–2Zr–1Mo–1V alloy by artificial neural network. *Mater. Des.* **2013**, *50*, 51–61.

7. Lin, Y.C.; Chen, X.-M. A critical review of experimental results and constitutive descriptions for metals and alloys in hot working. *Mater. Des.* **2011**, *32*, 1733–1759.

8. Huang, Y.-C.; Lin, Y.C.; Deng, J.; Liu, G.; Chen, M.-S. Hot tensile deformation behaviors and constitutive model of 42CrMo steel. *Mater. Des.* **2014**, *53*, 349–356.

9. Lin, Y.-C.; Chen, M.-S.; Zhang, J. Modeling of flow stress of 42CrMo steel under hot compression. *Mater. Sci. Eng. A* **2009**, *499*, 88–92.

10. Ji, G.; Li, F.; Li, Q.; Li, H.; Li, Z. A comparative study on Arrhenius-type constitutive model and artificial neural network model to predict high-temperature deformation behaviour in Aermet100 steel. *Mater. Sci. Eng. A* **2011**, *528*, 4774–4782.

11. Peng, W.; Zeng, W.; Wang, Q.; Yu, H. Comparative study on constitutive relationship of As-cast Ti60 titanium alloy during hot deformation based on Arrhenius-type and artificial neural network models. *Mater. Des.* **2013**, *51*, 95–104.

12. Wu, H.; Wen, S.P.; Huang, H.; Wu, X.L.; Gao, K.Y.; Wang, W.; Nie, Z.R. Hot deformation behavior and constitutive equation of a new type Al–Zn–Mg–Er–Zr alloy during isothermal compression. *Mater. Sci. Eng. A* **2016**, *651*, 415–424.

13. Ji, G.; Li, F.; Li, Q.; Li, H.; Li, Z. Prediction of the hot deformation behavior for Aermet100 steel using an artificial neural network. *Comput. Mater. Sci.* **2010**, *48*, 626–632.

14. Sabokpa, O.; Zarei-Hanzaki, A.; Abedi, H.R.; Haghdadi, N. Artificial neural network modeling to predict the high temperature flow behavior of an AZ81 magnesium alloy. *Mater. Des.* **2012**, *39*, 390–396.

15. Haghdadi, N.; Zarei-Hanzaki, A.; Khalesian, A.R.; Abedi, H.R. Artificial neural network modeling to predict the hot deformation behavior of an A356 aluminum alloy. *Mater. Des.* **2013**, *49*, 386–391.

16. Lu, Z.; Pan, Q.; Liu, X.; Qin, Y.; He, Y.; Cao, S. Artificial neural network prediction to the hot compressive deformation behavior of Al–Cu–Mg–Ag heat-resistant aluminum alloy. *Mech. Res. Commun.* **2011**, *38*, 192–197.

17. Xiao, X.; Liu, G.Q.; Hu, B.F.; Zheng, X.; Wang, L.N.; Chen, S.J.; Ullah, A. A comparative study on Arrhenius-type constitutive equations and artificial neural network model to predict high-temperature deformation behaviour in 12Cr3WV steel. *Comput. Mater. Sci.* **2012**, *62*, 227–234.

18. Han, Y.; Qiao, G.; Sun, J.; Zou, D. A comparative study on constitutive relationship of As-cast 904L austenitic stainless steel during hot deformation based on Arrhenius-type and artificial neural network models. *Comput. Mater. Sci.* **2013**, *67*, 93–103.

19. Li, H.-Y.; Wang, X.-F.; Wei, D.-D.; Hu, J.-D.; Li, Y.-H. A comparative study on modified Zerilli-Armstrong, Arrhenius-type and artificial neural network models to predict high-temperature deformation behavior in T24 steel. *Mater. Sci. Eng. A* **2012**, *536*, 216–222.

20. Gupta, A.K.; Singh, S.K.; Reddy, S.; Hariharan, G. Prediction of flow stress in dynamic strain aging regime of austenitic stainless steel 316 using artificial neural network. *Mater. Des.* **2012**, *35*, 589–595.

21. Bahrami, A.; Mousavi-Anijdan, S.H.; Madaah-Hosseini, H.R.; Shafyei, A.; Narimani, R. Effective parameters modeling in compression of an austenitic stainless steel using artificial neural network. *Comput. Mater. Sci.* **2005**, *34*, 335–341.

22. Zhu, Y.; Zeng, W.; Sun, Y.; Feng, F.; Zhou, Y. Artificial neural network approach to predict the flow stress in the isothermal compression of As-cast TC21 titanium alloy. *Comput. Mater. Sci.* **2011**, *50*, 1785–1790.

23. Serajzadeh, S. Prediction of temperature distribution and required energy in hot forging process by coupling neural networks and finite element analysis. *Mater. Lett.* **2007**, *61*, 3296–3300.

24. Phaniraj, M.P.; Lahiri, A.K. The applicability of neural network model to predict flow stress for carbon steels. *J. Mater. Process. Technol.* **2003**, *141*, 219–227.

25. Srinivasa, N.; Prasad, Y.V.R.K. Hot working characteristics of nimonic 75, 80A and 90 superalloys: A comparison using processing maps. *J. Mater. Process. Technol.* **1995**, *51*, 171–192.

26. Sheikh, H.; Serajzadeh, S. Estimation of flow stress behavior of AA5083 using artificial neural networks with regard to dynamic strain ageing effect. *J. Mater. Process. Technol.* **2008**, *196*, 115–119.

27. Sellars, C.M.; Mctegart, W.J. On the mechanism of hot deformation. *Acta Metall.* **1966**, *14*, 1136–1138.

28. Zener, C.; Hollomon, J.H. Effect of strain rate upon plastic flow of steel. *J. Appl. Phys.* **1944**, *15*, 22–32.

29. Liu, Y.; Yao, Z.; Ning, Y.; Nan, Y.; Guo, H.; Qin, C.; Shi, Z. The flow behavior and constitutive equation in isothermal compression of FGH4096-GH4133B dual alloy. *Mater. Des.* **2014**, *63*, 829–837.

30. Samantaray, D.; Mandal, S.; Bhaduri, A.K. A comparative study on Johnson Cook, modified Zerilli-Armstrong and Arrhenius-type constitutive models to predict elevated temperature flow behaviour in modified 9Cr–1Mo steel. *Comput. Mater. Sci.* **2009**, *47*, 568–576.

31. Cai, J.; Wang, K.; Zhai, P.; Li, F.; Yang, J. A modified Johnson-Cook constitutive equation to predict hot deformation behavior of Ti–6Al–4V alloy. *J. Mater. Eng. Perform.* **2014**, *24*, 32–44.

32. Wang, W.-T.; Guo, X.-Z.; Huang, B.; Tao, J.; Li, H.-G.; Pei, W.-J. The flow behaviors of clam steel at high temperature. *Mater. Sci. Eng. A* **2014**, *599*, 134–140.

Classifying Four Carbon Fiber Fabrics via Machine Learning: A Comparative Study Using ANNs and SVM

Min Zhao, Zijun Li and Wanfei He

Abstract: Carbon fiber fabrics are important engineering materials. However, it is confusing to classify different carbon fiber fabrics, leading to risks in engineering processes. Here, a classification method for four types of carbon fiber fabrics is proposed using artificial neural networks (ANNs) and support vector machine (SVM) based on 229 experimental data groups. Sample width, breaking strength and breaking tenacity were set as independent variables. Quantified numbers for the four carbon fiber fabrics were set as dependent variables. Results show that a multilayer feed-forward neural network with 21 hidden nodes (MLFN-21) has the best performance for classification, with the lowest root mean square error (RMSE) in the testing set.

Reprinted from *Appl. Sci.* Cite as: Zhao, M.; Li, Z.; He, W. Classifying Four Carbon Fiber Fabrics via Machine Learning: A Comparative Study Using ANNs and SVM. *Appl. Sci.* **2016**, *6*, 209.

1. Introduction

Carbon fiber is a new engineering material, which has become popular in aerospace, missile and rocket development [1–5]. In recent years, these materials have been developed rapidly in the field of civil construction, including in architecture and sports [6,7]. The resin composite sheet of carbon fiber fabrics is an enhanced product with unidirectional (UD) carbon fiber, which is frequently used because of its excellent mechanical properties and easy repairability [8]. Due to these superior properties, it is currently becoming a promising material in construction industries.

However, because the appearances of different carbon fiber fabrics have no significant difference, different types of carbon fabric fibers are very easy to confuse during productions [9,10], causing a large number of issues related to applications and construction. Also, there is a huge potential risk of compromised security in engineering processes using the resin composite sheet of carbon fiber fabrics without an exact classification [9]. So far, a direct classification approach for different fabrics is to measure the density [10]. However, this measurement is complicated and has high requirements with relevant instruments. Thus, there is still no study that reports an effective solution.

To classify different types of resin composite sheets of carbon fiber fabrics in a simpler way, we should firstly note that differences in density may lead to differences in tensile strength and relevant properties, which can be easily measured by fabric strength machines. Therefore, we can rationally assume that measuring the fabric of carbon fiber using a fabric strength machine can help us obtain a method for classification. Then the question becomes simpler: how do we find out the relationship between the properties acquired from the fabric strength machine and the classification result? Theoretical studies have offered several equations that describe relevant testing processes. Nevertheless, there is still a lack of an available method that can quantify different carbon fiber fabrics. Here, we successfully classify four different carbon fiber fabrics using a simple, defined quantification method with the strong classification capacity of machine learning techniques. Knowledge-based machine learning models were developed after the training process based on a large number of experimental data groups. To acquire enough experimental results, tensile stress and strain performances of four different types of carbon fiber fabrics were tested. All experimental data were measured from 231 samples in four different sample sizes. Based on the experimental data, novel machine learning techniques including artificial neural networks (ANNs) and support vector machine (SVM) were developed, respectively, for the classification of the four types of carbon fiber fabrics. This study, as an application research, aims at using user-friendly modeling techniques to help people classify different fabrics quickly based on the experimental data of tensile tests in research and practical applications. Therefore, the requirements of tensile strength can influence the selection of carbon fabrics by using our modeling techniques in practical applications.

2. Materials and Methods

2.1. Experimental

2.1.1. Preparation of a Resin Composite Sheet of Carbon Fiber Fabrics

To acquire an experimental database for model training, four typical carbon fiber fabrics were used during the experiments (Table 1). It can be apparently seen that the significant difference among the four fabrics is the density, ranging from 24 to 27.

To prepare resin composite materials, carbon fiber fabrics were impregnated with epoxy resin for 72 h. Specifications of epoxy resin were in accordance with the practical applications of the four kinds of carbon fiber fabrics respectively [9]. Afterwards, 30 cm of the carbon fiber fabric was extracted from the samples. The glass pane, polyester resin sheet and related tools were cleaned by absolute ethyl alcohol and dried. A polyester resin sheet was placed on a 10-cm-thick glass plane. Four-fifths of uniform epoxy resin was poured out and shaved to the polyester resin sheet. Carbon fiber laminates were placed on the uniform epoxy resin and rolled by a metal drum. The fiber should be kept being straight and epoxy resin should be kept transferring from the bottom to the surface of the fiber. Then the remaining one-fifth of epoxy resin was poured out to the surface of fiber uniformly. Afterwards, the samples were covered with polyester resin sheets. Bubbles and redundant resin were shaved away by a blade. A 5-mm-thick glass sheet was covered and all the samples were dried for seven days.

Table 1. Specifications of the four carbon fiber fabrics.

No.	Specification of Carbon Fiber Multifilament [a] (K)	Mean Fabric Density [b]	Grammes per Square Meter (g/m^2)
1	12	25	200
2	12	24	200
3	10	27	200
4	18	25.7	300

[a] Carbon fiber multifilament consists of a certain number of monofilament yarns; [b] Fabric density: number of carbon fiber multifilament in every 10 cm length of carbon fiber fabric.

2.1.2. Sampling

In order to avoid data distortion caused by uneven fabric of carbon fiber itself and to ensure the randomness of samples, the central fabrics of carbon fiber were selected randomly from the whole width of 2~2.5 m. Then 30 cm of the fabric being perpendicular to the edge was cut from the distance of the first 5 cm of the end of the fabric. The fabric defect and the joints of the carbon fiber were avoided.

After being stuck during impregnations, the end of the reinforced sheet of fabrics could be drawn by testing machines. The shapes of all tested samples were long rectangular (Figure 1).

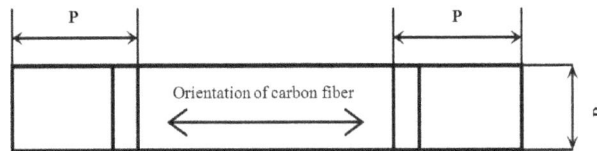

Figure 1. Schematic diagram of a tested sample. *B*: Sample wide; *P*: Length of carbon fiber reinforced sheet.

The lengths of the tested samples referred to the standards of both China and Japan [11–13], which were 230 mm. The widths of samples were designed as four specifications at the interval of 5 mm, including 15, 20, 25 and 30 mm.

2.1.3. Measurement

All samples were tested by the electronic universal testing machine (Shimadzu Co. Ltd., Kyoto, Japan, AG-10TA) for acquiring the breaking strengths and breaking tenacities. The environmental temperature was 14 °C and the humidity was 76%. The stretching speed was 5 m/min, which is in accordance with the standard GB/T1447–1983 [13].

2.2. Machine Learning Models

2.2.1. ANNs

ANNs are statistical learning tools for predictions and classifications in practical applications [14–19], which were invented from the inspiration of human brains. In an ANN, neurons in one layer are connected with all neurons in the next layer. Inter-connected neurons can tune the weight values combining the inputs in order to approximate the actual outputs. Therefore, ANNs are able to classify different objects with the same types of independent variables. Figure 2 is a schematic structure of a typical ANN for the classification of four carbon fiber fabrics, which contains the input, hidden and output layers.

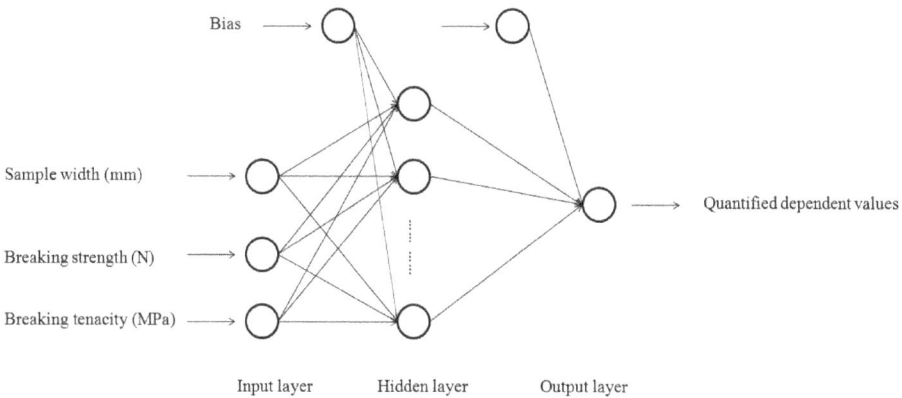

Figure 2. Schematic structure of an ANN (artificial neural network) for classifying the four carbon fiber fabrics.

2.2.2. SVM

SVM is a powerful machine learning method based on the statistical learning theory [20–22]. With limited information of samples between the complexity and learning ability of models, this theory is capacity of global optimization. In the basic principle of SVM, the target of a SVM is to find the optimal hyper-plane, a plane that separates all samples with the maximum margin [16,17,19,20]. This plane not only helps improve the predictive ability of the model, but also helps reduce the error which occurs occasionally when classifying. Figure 3 is the main structure of a typical SVM [16,22]. The letter "*K*" represents kernels. Small subsets extracted from the training data by relevant algorithm help develop the SVM. For applications, choosing suitable kernels and parameters is of great crucial to get a good classification result. With the development of programming, the use of software packages is able to help us solve this problem in a relatively reliable way [21].

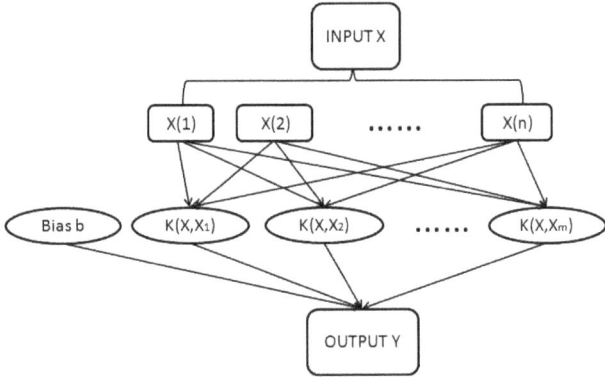

Figure 3. Schematic structure of an SVM (support vector machine) [22].

3. Results and Discussion

3.1. Model Development

Since machine learning models are developed based on the existing database acquired from experiments, here all the experimental results were used for the model training and testing. Statistical descriptions of the experimental results are shown in Table 2.

Table 2. Statistical descriptions of the experimental results.

Statistical Item	Sample Width (mm)	Breaking Strength (N)	Breaking Tenacity (MPa)
Minimum	15	3867.3	1621
Maximum	30	22,618	5988
Range	15	1875.7	4367
Average	239	110,662	3087.61
Standard deviation	5.39	4011.17	1020.92

Due to the powerful learning capacity of ANNs and SVM, we can define the corresponding quantified values by ourselves to classify the four different carbon fiber fabrics, respectively. Here, we define the samples numbers 1–4 as 200, 400, 600 and 800, respectively. Sample width, breaking strength and breaking tenacity were set as the independent variables, while the defined quantified classification for the four kinds of typical samples were set as the dependent values. Then, 85% of the data groups were set as the training set, while the remaining 15% were set as the testing set. ANNs were developed by NeuralTools® software (trial version, Palisade Corporation, Ithaca, NY, USA) [22,23]. A general regression neural network (GRNN) [24,25] and multilayer feed-forward neural networks (MLFNs) [26,27] were used from the software. The SVM model was developed by Matlab software (Libsvm package [21]). The computer for model development was a Lenovo G480 (laptop). To find out the best results of the MLFNs, the nodes in the hidden layer were set from 2 to 50. To measure the performance of different machine learning models, root mean square error (RMSE) and required training time were used as indicators that could help us define the most suitable model. Model development results are shown in Table 3. Results show that the MLFN with 21 nodes (MLFN-21) has the lowest RMSE for the testing process (36.03), while the SVM and other ANNs have comparatively higher RMSEs and lower classification accuracies. The change regulation of MLFNs with different numbers of nodes (Figure 4) also shows that with the increase of the node numbers, the required training times of the MLFNs gradually increase with a fluctuation. Though the training time of MLFN-21 is slightly longer than those of SVM, GRNN and other MLFNs with lower numbers of nodes, it is still acceptable because the training time will decrease with a high-performance computer. Therefore, the MLFN-21 can be rationally considered as the best model for classifying the four different carbon fiber fabrics in our experiments.

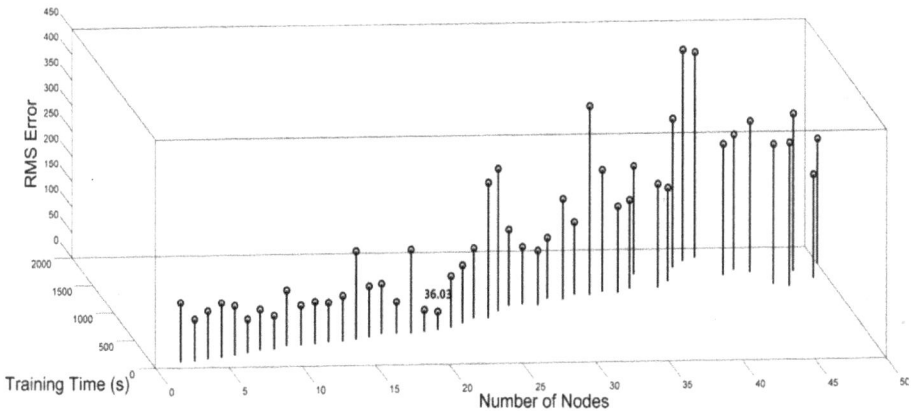

Figure 4. Change regulation of RMSEs (root mean square errors) in MLFNs (multilayer feed-forward neural networks) with different number of nodes.

Table 3. Best net search in different models. SVM: support vector machine; GRNN: general regression neural network; MLFN: multilayer feed-forward neural network.

Model	RMSE (for Testing)	Required Training Time
SVM	157.18	0:00:01
GRNN	154.79	0:00:01
MLFN 2 Nodes	116.51	0:01:45
MLFN 3 Nodes	81.92	0:01:58
MLFN 4 Nodes	94.38	0:02:30
MLFN 5 Nodes	107.45	0:02:52
MLFN 6 Nodes	98.24	0:03:27
MLFN 7 Nodes	65.92	0:04:11
MLFN 8 Nodes	79.64	0:05:01
MLFN 9 Nodes	66.09	0:05:09
MLFN 10 Nodes	109.95	0:06:02
MLFN 11 Nodes	78.88	0:06:13
MLFN 12 Nodes	83.69	0:06:28
MLFN 13 Nodes	78.48	0:06:59
MLFN 14 Nodes	90.54	0:07:09
MLFN 15 Nodes	174.10	0:07:42
MLFN 16 Nodes	100.59	0:08:21
MLFN 17 Nodes	98.58	0:09:19
MLFN 18 Nodes	62.81	0:09:17
MLFN 19 Nodes	165.28	0:09:15
MLFN 20 Nodes	43.10	0:09:40
MLFN 21 Nodes	36.03	0:10:11
MLFN 22 Nodes	101.73	0:10:51
MLFN 23 Nodes	114.28	0:12:06
MLFN 24 Nodes	138.72	0:13:26
MLFN 25 Nodes	266.98	0:13:27
MLFN 50 Nodes	246.41	0:28:23

3.2. Model Analysis

To analyze the performance of MLFN-21, its non-linear fitting process should be firstly discussed. In the model training process (Figure 5), the predicted values are generally close to the actual values (Figure 5a). Residual values are generally close to zero except for several discrete points (Figure 5b,c). The results of the training process show that the non-linear fitting process of MLFN-21 for classifying the four carbon fiber fabrics is decent. In terms of the model testing process (Figure 6), the predicted values in the testing set are very close to the actual values (Figure 6a), with comparatively low residual values (Figure 6b,c). The results of the testing process show that the MLFN-21 has a strong capacity for classifying the four carbon fiber fabrics, with the use of the four quantified dependent values: 200, 400, 600 and 800.

To evaluate the reproducibility of the ANNs, it should be firstly noted that the initial values of the weights are chosen randomly at the beginning of training. Then the initial weights are tuned according to the errors between the actual and desired outputs. Therefore, the training results of each single model development with the same component of the training and testing sets are always different, which is reflected in the fluctuation of RMSEs in the testing set. To prove that an ANN has good reproducibility, repeated experiments should be done in order to evaluate whether the RMSEs in the dependent training process have a stable fluctuation. To test the reproducibility of the MLFN-21 of this study, we repeated the training 100 times using the same component of the training and testing sets (Figure 7). It shows that MLFN-21 has a very stable fluctuation in the RMSE in testing during all repeated experiments. All RMSEs in testing are in the range between 21.1 and 41. The repeated experiments show that MLFN-21 has very good reproducibility for classifying the four carbon fiber fabrics.

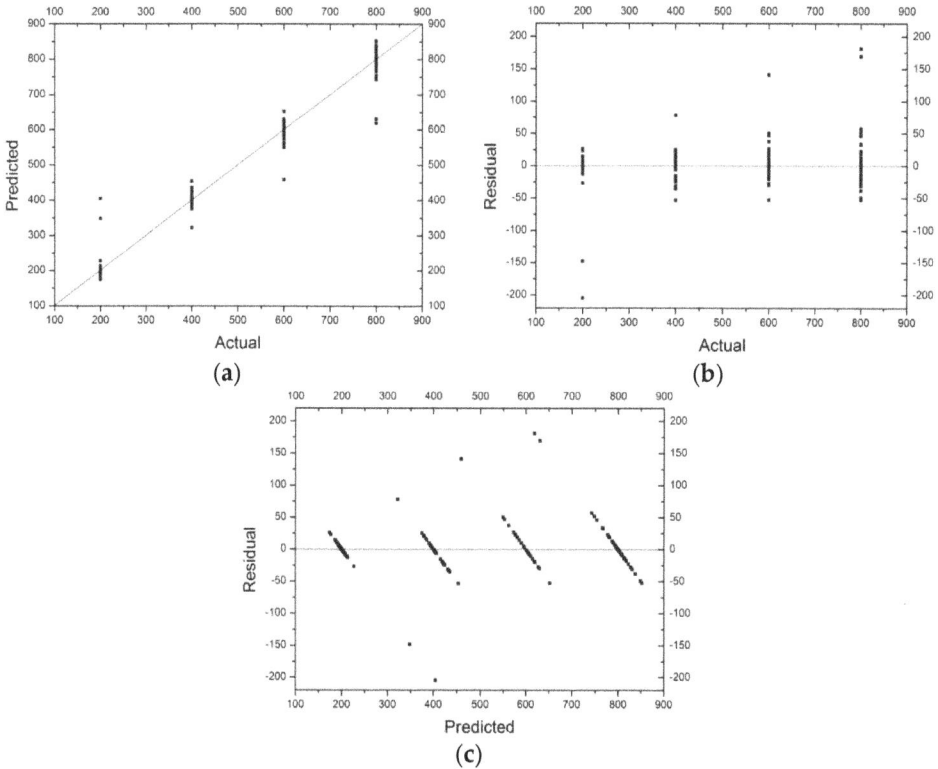

Figure 5. Training results of MLFN-21. (**a**) Predicted values versus actual values; (**b**) residual values versus actual values; (**c**) residual values versus predicted values.

101

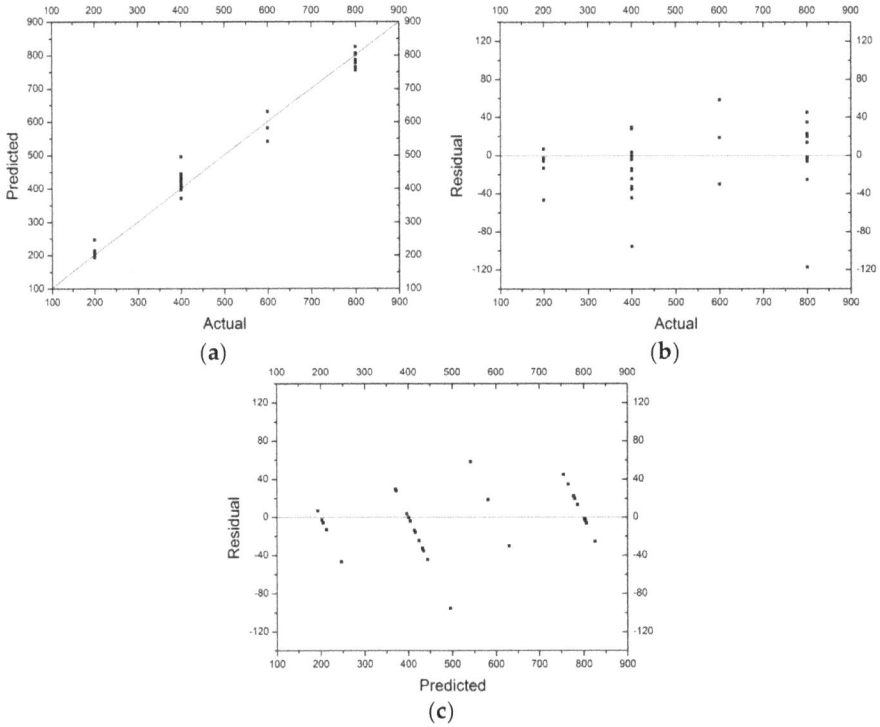

Figure 6. Testing results of MLFN-21. (**a**) Predicted values versus actual values; (**b**) residual values versus actual values; (**c**) residual values versus predicted values.

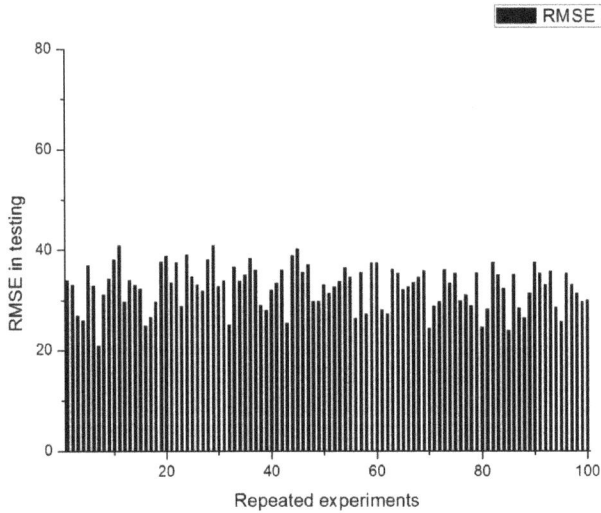

Figure 7. RMSEs of repeated experiments for developing the MLFN-21.

4. Conclusions

Here, ANNs and SVM are developed for classifying four types of carbon fiber fabrics. Results show that using four different numbers to represent the outputs of the four different fabrics with the use of the MLFN-21 model can help us classify different carbon fiber fabrics in real applications. With the inputs of sample width, breaking strength and breaking tenacity, this study successfully shows that machine learning methods, such as the MLFN-21, can effectively help us classify different carbon fiber fabrics based on the training of experimental data. It can be seen that ANNs are powerful tools to make good classification results due to their strong non-linear data-learning capacity. Also, it should be noted that in this study, the most important thing we were concerned with was the classification of carbon fiber fabrics, not the production. For classification, according to the principle, the ANN model is a "black box" non-linear training model. Thus, the determination of weights between different neuron layers was trained according to the iterations, not the exact correlations between independent and dependent variables. Therefore, it is uncertain to say which production factor will greatly affect the results in this study. Further research can be undertaken to develop a wider model for the determination of the correlation between the production factor and the types of carbon fabric fibers.

Author Contributions: Zijun Li did the experimental and modeling studies; Wanfei He and Min Zhao analyzed the results; all authors took part in the paper writing.

Conflicts of Interest: The authors declare no conflict of interest.

Abbreviations

The following abbreviations are used in this manuscript:

ANNs	artificial neural networks
SVM	support vector machine
GRNN	general regression neural network
MLFN	multilayer feed-forward neural network
RMSE	root mean square error

References

1. Motoyuki, S. Activated carbon fiber: Fundamentals and applications. *Carbon* **1994**, *32*, 577–586.
2. Chen, W.C. Some experimental investigations in the drilling of carbon fiber-reinforced plastic (CFRP) composite laminates. *Int. J. Mach. Tool. Manuf.* **1997**, *37*, 1097–1108.
3. Kumar, S.; Doshi, H.; Srinivasarao, M.; Park, J.; Schiraldi, D. Fibers from polypropylene/nano carbon fiber composites. *Polymer* **2002**, *43*, 1701–1703.

4. Bekyarova, E.; Thostenson, E.T.; Yu, A.; Kim, H.; Gao, J.; Tang, J.; Hahn, H.T.; Chou, T.W.; Itkis, M.E.; Haddon, R.C. Multiscale carbon nanotube-carbon fiber reinforcement for advanced epoxy composites. *Langmuir* **2007**, *23*, 3970–3974.

5. Sidoli, G.E.; King, P.A.; Setchell, D.J. An in vitro evaluation of a carbon fiber-based post and core system. *J. Prosthet. Dent.* **1997**, *78*, 5–9.

6. Norris, T.; Hamid, S.; Mohammad, R.E. Shear and flexural strengthening of R/C beams with carbon fiber sheets. *J. Struct. Eng. ASCE* **1997**, *123*, 903–911.

7. Seible, F.; Nigel Priestley, M.J.; Hegemier, G.A.; Innamorato, D. Seismic retrofit of RC columns with continuous carbon fiber jackets. *J. Compos. Constr.* **1997**, *1*, 52–62.

8. Gay, D. *Composite Materials: Design and Applications*; CRC Press: Boca Raton, FL, USA, 2014.

9. Luo, L.L.; Zhao, M.; Xu, C.X.; Zhou, F.X. A research on test methods for carbon fiber sheet. *Build. Sci. Res. Sichuan*; **2004**, *3*, 98–101.

10. Zhao, Q.; Zhao, M.; Xu, C.X. Tensile stress test for carbon fiber/epoxy composite sheet. *Sichuan Text. Technol.* **2004**, *4*, 40–41.

11. Research Subcommittee on Test Method and Specifications for Concrete Concrete Committee. *JSCE-E 542-2000 Tensile Strength Test Methods for Carbon Fiber Sheet*; Japan Society of Civil Engineers: Tokyo, Japan, 2000.

12. National Technical Committee on Fiber Reinforced Plastic of Standardization Administration of China. *GB-3354-82 Tensile Performance Test Methods for Unidirectional (UD) Reinforced Composite Fiber*; Standards Press of China: Beijing, China, 1998.

13. National Technical Committee on Fiber Reinforced Plastic of Standardization Administration of China. *GB-1447-83 Tensile Performance Test Methods for Glass Fiber Reinforced Plastic (GFRP)*; Standards Press of China: Beijing, China, 1983.

14. Hopfield, J.J. Artificial neural networks. *IEEE Circuit. Devices* **1988**, *4*, 3–10.

15. Drew, P.J.; Monson, R.T. Artificial neural networks. *Surgery* **2000**, *127*, 3–11.

16. Chen, F.D.; Li, H.; Xu, Z.H.; Hou, S.X.; Yang, D.Z. User-friendly optimization approach of fed-batch fermentation conditions for the production of iturin A using artificial neural networks and support vector machine. *Electron. J. Biotechnol.* **2015**, *18*, 273–280.

17. Li, H.; Chen, F.D.; Cheng, K.W.; Zhao, Z.Z.; Yang, D.Z. Prediction of Zeta Potential of Decomposed Peat via Machine Learning: Comparative Study of Support Vector Machine and Artificial Neural Networks. *Int. J. Electrochem. Sci.* **2015**, *10*, 6044–6056.

18. Li, H.; Yang, D.Z.; Chen, F.D.; Zhou, Y.B. Application of Artificial Neural Networks in predicting abrasion resistance of solution polymerized styrene-butadiene rubber based composites. In Proceedings of the 2014 IEEE Workshop on Electronics, Computer and Applications, Ottawa, ON, Canada, 8–9 May 2014; pp. 581–584.

19. Yang, D.Z.; Li, H.; Cao, C.C.; Chen, F.D.; Zhou, Y.B.; Xiu, Z.L. Analysis of the Oil Content of Rapeseed Using Artificial Neural Networks Based on Near Infrared Spectral Data. *J. Spectrosc.* **2014**, *2014*.

20. Suykens, J.A.K.; Vandewalle, J. Least squares support vector machine classifiers. *Neural Process. Lett.* **1999**, *9*, 293–300.

21. Chang, C.C.; Lin, C.J. LIBSVM: A library for support vector machines. *ACM TIST* **2011**, *2*.

22. Li, H.; Tang, X.; Wang, R.; Lin, F.; Liu, Z.; Cheng, K. Comparative Study on Theoretical and Machine Learning Methods for Acquiring Compressed Liquid Densities of 1,1,1,2,3,3,3-Heptafluoropropane (R227ea) via Song and Mason Equation, Support Vector Machine, and Artificial Neural Networks. *Appl. Sci.* **2016**, *6*.

23. Mohamed, M.; Nedret, B. Recognition of western style musical genres using machine learning techniques. *Expert Syst. Appl.* **2009**, *36*, 11378–11389.

24. Specht, D.F. A general regression neural network. *IEEE Neural Netw.* **1991**, *26*, 568–576.

25. Leung, M.T.; Chen, A.S.; Daouk, H. Forecasting exchange rates using general regression neural networks. *Comput. Oper. Res.* **2000**, *27*, 1093–1110.

26. Svozil, D.; Kvasnicka, V.; Pospichal, J. Introduction to multi-layer feed-forward neural networks. *Chemomet. Intel. Lab.* **1997**, *39*, 43–62.

27. Liu, Z.; Li, H.; Zhang, X.; Jin, G.; Cheng, K. Novel Method for Measuring the Heat Collection Rate and Heat Loss Coefficient of Water-in-Glass Evacuated Tube Solar Water Heaters Based on Artificial Neural Networks and Support Vector Machine. *Energies* **2015**, *8*, 8814–8834.

A Selective Dynamic Sampling Back-Propagation Approach for Handling the Two-Class Imbalance Problem

Roberto Alejo, Juan Monroy-de-Jesús, Juan H. Pacheco-Sánchez,
Erika López-González and Juan A. Antonio-Velázquez

Abstract: In this work, we developed a Selective Dynamic Sampling Approach (SDSA) to deal with the class imbalance problem. It is based on the idea of using only the most appropriate samples during the neural network training stage. The "average samples"are the best to train the neural network, they are neither hard, nor easy to learn, and they could improve the classifier performance. The experimental results show that the proposed method is a successful method to deal with the two-class imbalance problem. It is very competitive with respect to well-known over-sampling approaches and dynamic sampling approaches, even often outperforming the under-sampling and standard back-propagation methods. SDSA is a very simple method for automatically selecting the most appropriate samples (average samples) during the training of the back-propagation, and it is very efficient. In the training stage, SDSA uses significantly fewer samples than the popular over-sampling approaches and even than the standard back-propagation trained with the original dataset.

Reprinted from *Appl. Sci.* Cite as: Alejo, R.; Monroy-de-Jesús, J.; Pacheco-Sánchez, J.H.; López-González, E.; Antonio-Velázquez, J.A. A Selective Dynamic Sampling Back-Propagation Approach for Handling the Two-Class Imbalance Problem. *Appl. Sci.* **2016**, *6*, 200.

1. Introduction

In recent years, the class imbalance problem has been a hot topic in machine learning and data-mining [1,2]. It appears when the classifier is trained with a dataset where the number of samples in one class is lower than the samples in the other class, this and produces an important deterioration in the classifier performance [3,4].

The common methods handled with the class imbalance problem have been the re-sampling methods (under-sampling and over-sampling) [2,5,6], mainly due to the independence of the underlying classifier [7]. One of the most well-known over-sampling methods is the Synthetic Minority Over-sampling Technique (SMOTE). This generates artificial samples of the minority class by interpolating existing instances that lie close together [8]. The development of other samplings has been motivated: borderline-SMOTE, Adaptive Synthetic Sampling (ADASYN), SMOTE editing nearest neighbor, safe-level-SMOTE, Density-Based Synthetic Minority Over-sampling TEchnique (DBSMOTE), SMOTE + Tomek's Links [9], among others (see [1,7,10]).

An interest has been observed for finding the best samples to build the classifiers. For example, borderline-SMOTE has been proposed to over-sample only the minority samples near the class decision borderline [11]. Accordingly, in [12], the safe-level-SMOTE is proposed, to select minority class instances from the safe level region, and then, these samples are used to generate synthetic instances. ADASYN has been developed to generate more synthetic data from minority class samples that are harder to learn than those from minority class samples, which are easy to learn [13]. In a similar way, SPIDER approaches (framework that integrates a selective data pre-processing with the Ivotes ensemble method) over-sampling locally only for those minority class samples that are difficult to learn and includes a removing or relabeling process of noisy samples from the majority class [14,15]. The above discussed approaches have in common that they use the K nearest neighbors rule as the basis, and they are applied before the classifier training stage.

On the other hand, the under-sampling methods have shown effectiveness to deal with the class imbalance problem (see [7,8,10,16–19]). One of the most successful under-sampling methods has been the random under-sampling, which eliminates random samples from the original dataset (usually from the majority class) to decrease the class imbalance, however, this method loses effectiveness when removing significant samples [7]. Other important under-sampling methods including a heuristic mechanism are: the neighborhood cleaning rule, from Wilson editing [20], one-sided selection [21], Tomek links [22] and the Condensed Nearest Neighbor rule (CNN) [23]. Basically, the aim of the cleaning mechanism is: (i) to eliminate samples with a high likelihood of being noise or atypical samples or (ii) to eliminate redundant samples in CNN methods. In the same way as the above approaches, we apply these methods before the training process. They employ the K nearest neighbors rule (except the Tomek links methods) as the basis.

Another important alternative to face the class imbalance has been the Cost Sensitive (CS) approach, which has become one of the most relevant topics in machine learning research in recent years [24]. They consider the costs associated with misclassifying samples, i.e., CS methods use different cost matrices describing the costs for misclassifying any particular data sample [10]. The over- and under-sampling could be a special case of the CS techniques [25]. Anther CS method is threshold-moving, which moves the output threshold toward inexpensive classes, such that samples with higher costs become hard to misclassify. It is applied in the test phase and does not affect the training phase [24].

Ensemble learning is an effective method that has increasingly been adopted to combine multiple classifiers and class imbalance approaches to improve the classification performance [2,4,5]. In order to combine the multiple classifiers, it is common to use the hard and soft ensemble. The former uses binary votes, while the latter uses real-valued votes [26].

Recently, dynamic sampling methods have become an interesting way to deal with the class imbalance problem. They are attractive, because they automatically find the proper sampling amount for each class in the training stage (different from conventional strategies as over- and/or under-sampling techniques). In addition, some dynamic sampling methods also identify the "best samples" for classifier training. For example, Lin et al. [27] propose a dynamic sampling method with the ability to identify samples with a high probability to be misclassified. The idea is that the classifier trained with these samples may produce better classification results. Other methods that can be considered as dynamic sampling are: (i) the snowball method (proposed in [28] and used as a dynamic training method in [29,30]); (ii) the genetic dynamic training technique [31,32]; in it, the authors employ a genetic algorithm to find the best over-sampling ratio; (iii) the mean square error (MSE) dynamic over-sampling method [19], which is based on the MSE back-propagation for automatically identifying the over-sampling rate. Chawla et al. [33] present a WRAPPER paradigm (for which the search is guided by the classification goodness measure as score) to discover the amount of the under-sampling and over-sampling rate for a dataset. Debowski et al. [34] show a very similar work.

The dynamic sampling approaches are a special case of the sampling techniques. The main difference of these methods with respect to the conventional sampling strategies is in the time when they sample the data or when they select the examples to be sampled (see [19,27,28,31,32]).

In this paper, a Selective Dynamic Sampling Approach (SDSA) to deal with the two-class imbalance problem is presented. This method is useful to find automatically the appropriate sampling amount for each class through the selection of the "best samples" to train the multilayer perceptron with the back-propagation algorithm [35]. The proposed method was tested over thirty five real datasets and compared to some state-of-the-art class imbalance approaches.

2. Selective Dynamic Sampling Approach

Researchers in the class imbalance problem have shown their interest in finding the best samples to build the classifiers, for example eliminating those samples with a high probability to be noise or overlapped samples [18,36–40], or focusing on those close to the borderline decision [11,13,41] (the latter has been less explored).

In accordance with the above discussion, three categories of samples can be basically identified in the class imbalance literature:

- Noise and rare or outlier samples. The first ones are instances with error in their labels [7] or erroneous values in the features that describe them, and the last ones are the minority and rare samples located inside the majority class [42].
- Border or overlapped samples are those samples located where the decision boundary regions intersect [18,38].
- Safe samples are those with a high probability of being correctly labeled by the classifier, and they are surrounded by samples of the same class [42].

Nevertheless, those samples situated close to the borderline decision and far from the safe samples might be of interest; in other words, those that are neither hard nor easy to learn. These samples are called "average samples" [35].

In this section, a Selective Dynamic Sampling Approach (SDSA) to train the multilayer perceptron is presented. The aim of this proposal is to deal with the two-class imbalance problem, i.e., this method only works with two-class imbalanced datasets. This SDSA is based on a modification of the "stochastic" back-propagation algorithm and derived from the idea of using average samples to train Arificial Neural Networks (ANN), in order to try to improve the classifier performance. The proposed method consists of two steps, and it is described below:

1. Before training: The training dataset is balanced 100% through an effective over-sampling technique. In this work, we use the SMOTE [8] (SDSAS) and random over-sampling (SDSAO) [16].
2. During training: The proposed method selects the average samples to update the neural network weights. From the balanced training dataset, it chooses average samples to use in the neural network training. With the aim to identify the average samples, we propose the next function:

$$\gamma(\Delta^q) = exp(-\frac{||\Delta^q - \mu||^2}{2\sigma^2})$$ (1)

Variable Δ^q is the normalized difference amongst the real neural network outputs for a sample q,

$$\Delta^q = \frac{z_0^q}{\sqrt{(z_0^q - z_1^q)^2}} - \frac{z_1^q}{\sqrt{(z_0^q - z_1^q)^2}}$$ (2)

where z_0^q and z_1^q are respectively the real neural network outputs corresponding to a q sample. The ANN only has two neural network outputs (z_0^q and z_1^q), because it has been designed to work with datasets of two classes [43].

The Selective Dynamic Sampling Approach (SDSA) is detailed in Algorithm 1, where $t_j^{(q)}$ and $z_j^{(q)}$ are the desired and real neural network outputs for a sample q, respectively.

Algorithm 1 The Selective Dynamic Sampling Approach (SDSA) based on the stochastic back-propagation multilayer perceptron.

Input: X (input dataset), N (number of features in X), K (number of classes in X), Q (number of samples in X), M (number of middle neurodes), J (number output neurodes), I number of iterations and learning rate η.

Output: the weights $\mathbf{w} = (w_{11}, w_{21}, ..., w_{NM})$ $\mathbf{u} = (u_{11}, u_{21}, ..., w_{MJ})$.

INIT( ):

1: Read MLP file (X, N, M, J, Q, I and η);
2: Generate initial weights randomly between -0.5 and 0.5;

LEARNING( ):

3: **while** $i < I$ or $E > 0.001$ **do**

4:     $x^q \leftarrow$ randomly chose a sample from X

5:     **FORWARD**(x^q);

6:     $\Delta^q = (z_0^q / \sqrt{(z_0^q - z_1^q)^2}) - (z_1^q / \sqrt{(z_0^q - z_1^q)^2})$;

7:     $\gamma(\Delta^q) = exp(-||\Delta^q - \mu||^2 / 2\sigma^2)$;

8:     **if** **Random( )** $<= \gamma(\Delta^q)$ **then**

9:         **UPDATE**(x^q);

10:     **end if**

11:     $i++$;

12: **end while**

FORWARD(x^q):

13: **for** $m = 0$ to $m < M$ **do**

14:     **for** $n = 0$ to $n < N$ **do**

15:         $y_m \leftarrow y_m + x_n^q * w_{nm}$;

16:     **end for**

17:     $y_m = net(y_m)$;

18: **end for**

19: **for** $j = 0$ to $j < J$ **do**

20:     **for** $m = 0$ to $m < M$ **do**

21:         $z_j \leftarrow z_j + u_{mj} * y_m$;

22:     **end for**

23:     $z_j \leftarrow net(z_j)$;

24: **end for**

UPDATE(x^q):

25: **for** $m = 1$ to M **do**

26:     **for** $j = 1$ to J **do**

27:         $u_{mj}^{r+1} \leftarrow u_{mj}^r + \eta\{(t_j^{(q)} - z_j^{(q)})[z_j^{(q)}(1 - z_j^{(q)})]y_m^{(q)}\}$;

28:     **end for**

29:     **for** $n = 1$ to N **do**

30:         $w_{nm}^{r+1} \leftarrow$

        $w_{nm}^r + \eta\{\sum_{j=1,J}(t_j^{(q)} - z_j^{(q)})[z_j^{(q)}(1 - z_j^{(q)})]u_{mj}^{(r)}\}x_n[y_m^{(q)}(1 - y_m^{(q)})][x_n^{(q)}]$;

31:     **end for**

32: **end for**

2.1. Selecting μ Values

The appropriate selection of the variable μ is critical to select the average samples or other kind of samples (border or safe samples [42]). Variable μ is computed under the following consideration: the target ANN outputs (t_j) are usually codified in zero and one values [43]. For example, for a two-class problem (Class A and Class B), the desired ANN outputs are codified as $(1,0)$ and $(0,1)$ for Classes A and B, respectively. These values are the target ANN outputs (t_j), i.e., the desired final values emitted by the ANN after training. In accordance with this understanding, the expected μ values are:

- $\mu \approx 1.0$ for safe samples. It is expected that ANN classifies with a high accuracy level, i.e., it is expected that the real ANN outputs for all neurons (z_j) will be values close to $(1, 0)$ and $(0, 1)$ for Classes A and B, respectively. Whether we apply Equation (2), the expected value is 1.0, at which the γ function has its maximum value.
- $\mu \approx 0.0$ for border samples. It is expected that the classifier misclassifies. The expected ANN outputs for all neurons are values close to $(0.5, 0.5)$, then the Δ is approximately 0.0, at which the γ function has its maximum value for these samples.
- $\mu \approx 0.5$ for average samples. It is expected that ANN classifies correctly, but with less accuracy. In addition, the average samples are between safe $(\mu \approx 1.0)$ and border $(\mu \approx 0.0)$ samples.

The recommended μ values to select the average samples are those around 0.5. An independent validation set to find the most appropriate μ value is proposed to avoid any bias in the testing process.

For this independent validation, a minimal subset from the training data is used. Firstly, the ten-fold cross-validation for each dataset is applied (Section 5.1); next, only 10% of samples are randomly taken from each training fold (TF^{10}), then TF^{10} is split into two disjoints folds of the same size $(TF^5_{train}$ and TF^5_{test}, respectively). Next, the proposed method (SDSA) is applied over the TF^5_{train} and TF^5_{test} to find the best μ value. The tested values for μ were 0.25, 0.375, 0.5, 0.625 and 0.75. Finally, the μ value, for which the best Area Under the Curve (AUC) [44] rank was obtained, is chosen by SDSA on TF^{10}.

Note that this independent validation does not imply an important computational cost, because it only uses 10% of the training data to find the most appropriate μ value. This independent validation unbiased the performance on the testing data process, due to the test data not being used.

3. State-of-the-Art of the Class Imbalance Approaches

In the state-of-the-art class imbalance problem, the over- and under-sampling methods are very popular and successful approaches to deal with this problem (see [7,8,10,16–19]). Over-sampling replicates samples in the minority-class, and under-sampling eliminates samples from the majority-class, biasing the discrimination process to compensate for the class imbalance.

This section describes some well-known sampling approaches that have been effectively applied to deal with the class imbalance problem. These approaches are used with the aim to compare the classification performance of the proposed method with respect to the state-of-the-art of class imbalance approaches.

3.1. Under-Sampling Approaches

TL Tomek links are pairs of samples a and b from different classes, and there does not exist a sample c, such that $d(a, c) < d(a, b)$ or $d(b, c) < d(a, b)$, where d is the distance between pairs of samples [22]. Samples in TL are noisy or lie in the decision border. This method removes those majority class samples belonging to TL [9].

CNN The main goal of the condensed nearest neighbor algorithm is the reduction of the size of the stored dataset of training samples while trying to maintain (or even improve) generalization accuracy. In this method, every member of X (the original training dataset) must be closer to a member of S (the pruned set) of the same class than any other member of S from a different class [23].

CNNTL combines the CNN with TL [9].

NCL The Neighborhood Cleaning Rule uses the Editing Nearest Neighbor (ENN) rule, but only eliminates the majority class samples. ENN uses the $k - NN$ ($k > 1$) classifier to estimate the class label of every sample in the dataset and discards those samples whose class labels disagree with the class associated with the majority of the k neighbors [20].

OSS The One-Sided Selection method performs TL, then CNN on the training dataset [21].

RUS The Random Under-Sampling randomly eliminates samples from the majority class and biases the discrimination process to compensate for the class imbalance.

3.2. Over-Sampling Approaches

ADASYN is an extension of SMOTE, creating more samples in the vicinity of the boundary among the two classes than in the interior of the minority class [13].

ADOMS The Adjusting the Direction Of the synthetic Minority clasS method, setting the direction of the synthetic minority class samples, this works like SMOTE, but it generates synthetic examples along the first component of the main axis of the local data distribution [45].

ROS The Random Over-Sampling duplicates samples randomly from the minority class, biasing the discrimination process to compensate for the class imbalance.

SMOTE [8] generates artificial samples of the minority class by interpolating existing instances that lie close together. It finds the k intra-class nearest neighbors for each minority sample, and then, synthetic samples are generated in the direction of some or all of those nearest neighbors.

B-SMOTE Borderline-SMOTE [11] selects samples from the minority class that are on the borderline (of the minority decision region, in the feature space) and only performs SMOTE on those samples, instead of over-sampling all or taking a random subset.

SMOTE-ENN This technique consists of applying the SMOTE and then applying the ENN rule [9].

SMOTE-TL is the combination of SMOTE and TL [9].

SL -SMOTE Safe-Level SMOTE is based on the SMOTE, but it generates synthetic minority class samples positioned closer to the largest safe level; then, all synthetic samples are only generated in safe regions [12].

SPIDER-1 is an approach that combines a local over-sampling of those minority class samples that are difficult to learn with removing or relabeling noisy samples from the majority class [14].

SPIDER-2 The major difference between this method and SPIDER-1 is that it divides into two stages the pre-processing of the majority and minority class samples, i.e., first pre-processing the majority class samples and next the minority class samples (considering the changes introduced in the first stage) [15].

4. Dynamic Sampling Techniques to Train Artificial Neural Networks

Dynamic sampling techniques have become an interesting way to deal with the class imbalance problem on the Multilayer Perceptron (MLP) trained with stochastic back-propagation [19,27,28,31,32]. Different from conventional strategies as over- and/or under-sampling techniques, the dynamic sampling finds automatically in the training stage the properly sampling amount for each class for dealing with the class imbalance problem. In this section, we present some details and the main features of two dynamic sampling methods.

4.1. *Method 1. Dynamic Sampling*

The basic idea of the Dynamic Sampling (DyS) method, proposed in [27], is to design a simple DyS that dynamically selects samples during the training process. In this method, a pre-deletion of any sample to prevent information loss, to dynamically select the samples (hard to classify) to train the ANN and to make the best use of the dataset does not exist. According to this main idea, the general steps in each epoch can be described as follows.

1. Randomly fetch a sample q from the training dataset.
2. Estimate the probability p that the example should be used for the training.

$$p = \begin{cases} 1, & \text{if } \delta \leq 0 \\ exp(-\delta \cdot r_j / min\{r_i\}), & \text{otherwise,} \end{cases} \quad (3)$$

where $\delta = z_j^q - max_{i \neq c}\{z_i^q\}$. z_i^q is the i-th real ANN output of the sample q and j is the class label to which q belongs. $r_c = Q_c/Q$ is the class ratio; Q_c is the number of samples belonging to class c; and Q is the sample number.

3. Generate a uniform random real number μ between zero and one.
4. If $\mu < p$, then use the sample q to update the weights by the back-propagation rules.
5. Repeat Steps 1–4 on all samples of the training dataset in each training epoch.

In addition, the authors of the paper [27] use an over-sampling method based on a heuristic technique to avoid bias for the class imbalance problem. Beginning with the first epoch, the process consists of the samples of all classes, except the largest classes over-sampled to make the dataset balanced. As the training process goes on, the over-sampling ratio (ρ) is attenuated in each epoch (ep) by a heuristic technique (Equation (4)). It is calculated as:

$$\rho = (r_{max}/r_j)/ln(ep) \quad (4)$$

where ep (> 2) and max represent the largest majority class.

4.2. Method 2. Dynamic Over-Sampling

In [19], a Dynamic Over-Sampling (DOS) technique to deal with the class imbalance problem was proposed. The main idea of DOS is to balance the MSE on the training stage (when a multi-class imbalanced dataset is used) through an over-sampling technique. Basically, the DOS method consists of two steps:

1. *Before training*: The training dataset is balanced at 100% through an effective over-sampling technique. In this work, SMOTE [8] is utilized.
2. *During training*: The MSE by class E_j is used to determine the number of samples by class (or class ratio) in order to forward it to the ANN. The equation employed to obtain the class ratio is defined as:

$$ratio_j = \frac{E_{max}}{E_j} \times \frac{Q_j}{Q_{max}}; \text{ for } j = 1, 2, ..., J \quad (5)$$

114

where J is the number of classes in the dataset and *max* identifies the largest majority class. Equation (5) allows balancing the MSE by class, reducing the impact of the class imbalance problem on the ANN.

The DOS method only uses the necessary samples for dealing with the class imbalance problem and, in this way, to avoid getting a poor classifications performance as a result of training the ANN with imbalanced datasets.

5. Experimental Set-Up

In this section, the techniques, datasets and experimental framework used in this paper are to be described.

5.1. Database Description

Firstly, for the experimental stage, five real-world remote sensing databases are chosen: Cayo, Feltwell, Satimage, Segment and 92AV3C. The Cayo dataset comes from a particular region in the Gulf of Mexico [18]. The Feltwell dataset represents an agricultural area near the village of Feltwell (UK) [46]. The Satimage and Segment datasets are from the UCI (University of California, Irvine) Machine Learning Database Repository [47]. The 92AV3C dataset [48] corresponds to a hyperspectral image (145 × 145 pixels, 220 bands, 17 classes) taken over the Northwestern Indiana Indian Pines by the AVIRIS (Airborne Visible / Infrared Imaging Spectrometer) sensor. In this work, we employed a reduced version of this dataset with six classes (2, 3, 4, 6, 7 and 8) and thirty eight attributes as in [18].

The two-class imbalance problem is only studied. We decompose the multi-class problems into multiple two-class imbalanced problems. This proceeds as follows: one class (c_j) is taken from the original database (*DB*) to integrate the minority class (c^+), and the rest of classes were joined to shape the majority class (c^-). Then, we integrate the two-class database DB_j ($j = 1, 2, ..., J$, and J is the number of classes in *DB*). In other words, $DB_j = c^+ \cup c^-$. Therefore, for each database, J two-class imbalanced datasets were obtained. The main characteristics of the new produced benchmarking datasets are shown in Table 1. This table shows that the datasets used in this work have several class imbalance levels (see the class imbalance ratio), ranging from a low to a high class imbalance ratio (for example, see 92A3 and CAY4 datasets). In addition, the ten-fold cross-validation method was applied on all datasets shown in this table.

Table 1. A brief summary of the main characteristics of the new produced benchmarking dataset.

Dataset	# of Features	# of Minority Classes Samples	# of Majority Class Samples	Imbalance Ratio
CAY0	4	838	5181	6.18
CAY1	4	293	5726	19.54
CAY2	4	624	5395	8.65
CAY3	4	322	5697	17.69
CAY4	4	133	5886	44.26
CAY5	4	369	5650	15.31
CAY6	4	324	5695	17.58
CAY7	4	722	5297	7.34
CAY8	4	789	5230	6.63
CAY9	4	833	5186	6.23
CAY10	4	772	5247	6.80
FELT0	15	3531	7413	2.10
FELT1	15	2441	8503	3.48
FELT2	15	896	10,048	11.21
FELT3	15	2295	8649	3.77
FELT4	15	1781	9163	5.14
SAT0	36	1508	4927	3.27
SAT1	36	1533	4902	3.20
SAT2	36	703	5732	8.15
SAT3	36	1358	5077	3.74
SAT4	36	626	5809	9.28
SAT5	36	707	5728	8.10
SEG0	19	330	1140	3.45
SEG1	19	50	1420	28.40
SEG2	19	330	1140	3.45
SEG3	19	330	1140	3.45
SEG4	19	50	1420	28.40
SEG5	19	50	1420	28.40
SEG6	19	330	1140	3.45
92A0	38	190	4872	25.64
92A1	38	117	4945	42.26
92A2	38	1434	3628	2.53
92A3	38	2468	2594	1.05
92A4	38	747	4315	5.78
92A5	38	106	4956	46.75

5.2. Parameter Specification for the Algorithms Employed in the Experimentation

The stochastic back-propagation algorithm was used in this work (the source code of back-propagation algorithm and the approaches (dynamic sampling methods) and the datasets used in this work are available at Ref. [49]), and for each training process, the weights were ten times randomly initialized. The learning rate (η) was set to 0.1, and we established the stopping criterion at 500 epochs or if the MSE value is lower than 0.001. A single hidden layer was used. The number of neurons in the hidden layer was set to four for every experiment.

All sampling methods (except ENN, SPIDER-1 and SPIDER-2, which employ three) use five nearest neighbors (if applicable) and sampling the training dataset to reach to relative class distribution balance (if applicable). ADASYN and ADOMS use the Euclidean distance, and the rest of the methods employ the Heterogeneous Value

Difference Metric (HVDM) [50], if applicable. SPIDER-1 applies a weak amplification pre-processing option, and SPIDER-2 employs relabeling of noisy samples from the majority class and an amplification option. The sampling methods have been done using the KEEL [51].

In order to identify the most suitable value for the variable μ, an independent validation set to avoid any bias in the performance on the testing data is considered, meaning that the testing data for this validation are not used (see Section 2.1). Thereafter, the most appropriate value for the variable μ obtained for the datasets used in this work (Table 1) is 0.375. The results presented in this paper were obtained with $\mu = 0.375$. In addition, for this independent validation, only 200 epochs are used in the neural network training stage and about 8% of the samples of each dataset. This does not imply an important additional computational effort. The SDSAO and SDSAS methods are the proposed methods using ROS and SMOTE, respectively (see Section 4).

5.3. Classifier Performance and Significant Statistical Test

The Area Under the receiver operating characteristic Curve (AUC) [44] was used as the criteria of measure for the classifiers performance. It is one of the most widely-used and accepted techniques for the evaluation of binary classifiers in class imbalance domains [10].

Additionally, in order to strengthen the results analysis, a non-parametric statistical test is achieved. The Friedman test is a non-parametric method in which the first step is to rank the algorithms for each dataset separately; the best performing algorithm should have rank as 1, the second best rank as 2, etc. In case of ties, average ranks are computed. The Friedman test uses the average rankings to calculate the Friedman statistic, which can be computed as,

$$\chi_F^2 = \frac{12N}{K(K+1)} \left(\sum_j R_j^2 - \frac{K(K+1)^2}{4} \right) \tag{6}$$

K denotes the number of methods; N is the number of data sets; and R_j is the average rank of method j on all datasets.

On the other hand, Iman and Davenport [52] demonstrated that χ_F^2 has a conservative behavior. They proposed a better statistic (Equation (7)) distributed according to the F−distribution with $K - 1$ and $(K-1)(N-1)$ degrees of freedom,

$$F_F = \frac{(N-1)\chi_F^2}{N(K-1) - \chi_F^2} \tag{7}$$

In this work, the Friedman and Iman–Davenport tests are employed with the $\gamma = 0.05$ level of confidence, and KEEL software [51] is utilized.

In addition, when the null-hypothesis was rejected, a post-hoc test is used in order to find the particular pairwise method comparisons producing statistically-significant differences. The Holm–Shaffer post-hoc tests are applied in order to report any significant difference between individual methods. The Holm procedure rejects the hypotheses (H_i) one at a time until no further rejections can be done [53]. To accomplish this, the Holm method ordains the p-values from the smallest to the largest, i.e., $p_1 \leq p_2 \leq p_{k-1}$, corresponding to the hypothesis sequence $H_1, H_2, ..., H_{k-1}$. Then, the Holm procedure rejects H_1 to H_{i-1} if i is the smallest integer, such that $p_i \leq \alpha/(k-i)$. This procedure starts with the most significant p-value. As soon as a certain null-hypothesis cannot be rejected, all of the remaining hypotheses are retained, as well [54]. The Shaffer method follows a very similar procedure to that proposed by Holm, but instead of rejecting H_i if $p_i \leq \alpha/(k-i)$, it rejects H_i if $p_i \leq \alpha/t_i$, where t_i is the maximum number of hypotheses that can be true given that any $(i-1)$ hypotheses are false [55].

6. Experimental Results and Discussion

In order to assess the performance of the proposed methods (SDSAO and SDSAS), a set of experiments has been carried out, over thirty five two-class datasets (Table 1) with ten well-known over-sampling approaches (ADASYN, ADOMS, B-SMOTE, ROS, SMOTE, SMOTE-ENN, SMOTE-TL, SPIDER-1, SPIDER-2 and SL-SMOTE), six popular under-sampling methods (TL, CNN, CNNTL, NCL, OSS and RUS) (for more detail about these re-sampling techniques, see Section 3) and two dynamic sampling approaches (DyS and DOS).

This section is organized as follows: First, the AUC values are shown, and the Friedman ranks are used to analyze the classification results (Table 2). Second, a statistical test is presented in order to strengthen the results discussion (Figure 1). Finally, the relationship between the training dataset size and the tested methods performance is studied (Figure 2).

The results presented in Table 2 are the AUC values obtained in the classifying stage, and they are averaged values between ten folds and ten different initialization weights of the neural network (see Section 5).

In accordance with the averaged ranks shown in Table 2, all over-sampling methods and dynamic sampling approaches (SDSAO, SDSAS, DyS and DOS) can improve the standard back-propagation (BP) performance, and the worst approaches with respect to standard BP are the under-sampling techniques, except by RUS, NCL and TL, which show a better performance than the standard BP. This table also shows that only the ROS technique presents a better performance than the proposed methods. SDSAO and DyS show a slight advantage over SDSAS.

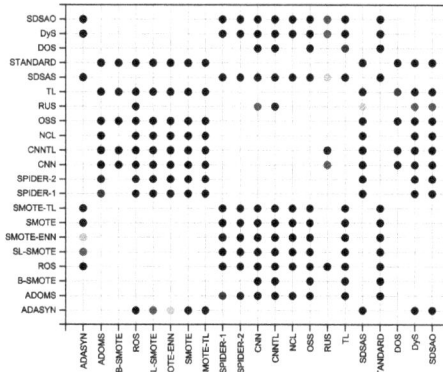

Figure 1. Results of the non-parametric statistical Holm and Shaffer post-hoc test. The fill circles mean that for these particular pairs of classifiers, the null hypothesis was rejected by both test. The color of the circles is the darkest at *p*-values close to zero, i.e., when the statistical difference is the most significant.

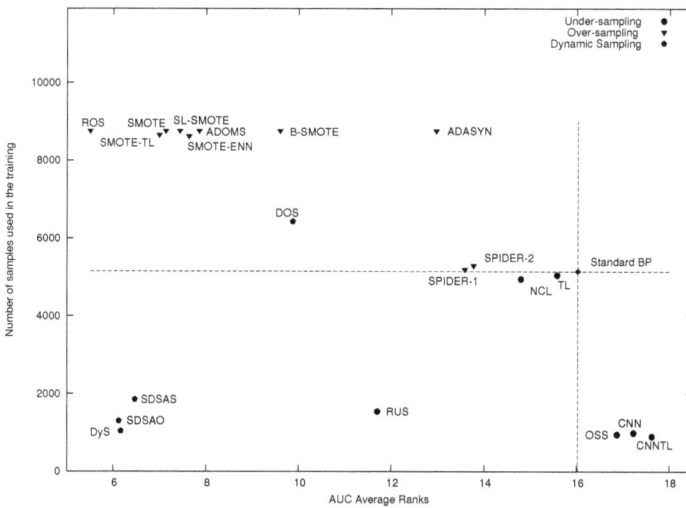

Figure 2. Number of samples used in the training process by the studied methods in contrast to the Area Under the receiver operating characteristic Curve (AUC) average ranks obtained in the classification test. The *x* axis represents the average ranks (the best performing method should have the rank of one or close to this value). We previously used the ten-fold cross-validation method. The number shown in the *y* axis corresponds to the average training fold size.

In addition, Table 2 indicates that the class Imbalance Ratio (IR) is not determinant in order to get high AUC values, for example CAY7, SAT2, SEG1, SEG5 and 92A5 datasets present high values of AUC no matter their IR; also in these datasets, most over-sampling methods and dynamic sampling approaches are very competitive.

Other datasets support this fact, i.e., IR is not critical in the classification performance, for example the SEG4 and SEG5 datasets have the same IR, but the classification performance (using the standard BP) is very different (values of AUC of 0.999 and 0.630, respectively). This confirms was was presented in other works, in that other features of the data might become a strong problem for the class imbalance [2]. For example: (i) the class overlapping or noisy data [39,42,56,57]; (ii) the small disjuncts; (iii) the lack of density and information in the training data [58]; (iv) the significance of the borderline instances [13,59] and their relationship with noisy samples; and (v) the possible differences in the data distribution for the training and testing data, also known as the dataset shift [7].

In order to strengthen the result analysis, a non-parametrical statistical and post-hoc tests are applied (see Section 5.3): Friedman and Iman–Davenport tests report that considering reduction performance distributed according to chi-square with 20 degrees of freedom, the Friedman statistic is set at 329.474, and the p-value computed by the Friedman test is 1.690×10^{-10}. However, considering reduction performance distributed according to the F-distribution with 20 and 680 degrees of freedom, the Iman and Davenport statistic is 30.233, and the p-value computed by their test is 2.588×10^{-80}. Then, the null hypothesis is rejected, i.e., the Friedman and Iman–Davenport tests indicate the existence of significant differences in the results. Due to these results, a post-hoc statistical analysis is required.

Figure 1 shows the results of the non-parametric statistical Holm and Shaffer post-hoc tests. The rows and columns constitute the studied methods; as a consequence, it represents all $C \times C$ pairwise classifier comparisons. The filled circles mean that for these particular pairwise methods (for $C_i \times C_j$; $i = 1, 2, ..., C$ and $i \neq j$), the null hypothesis was reject by the Holm–Shaffer post-hoc tests. Therefore, the color of circles is the darkest when the p-values are close to zero; this means that the statistical difference is significant.

Table 2. Back-propagation classification performance using the Area Under the receiver operating characteristic Curve (AUC). The results represent the averaged values between ten folds and the initialization of ten different weights of the neural network. The best values are underlined in order to highlight them. ROS, Random Over-Sampling; SDSAO, Selective Dynamic Sampling Approach using ROS; DyS, Dynamic Sampling; SDSAS, Selective Dynamic Sampling Approach applying SMOTE; SMOTE-TL, SMOTE and TL;SMOTE, Synthetic Minority Over-sampling Technique; SL-SMOTE, Safe-Level SMOTE; SMOTE-ENN, SMOTE and Editing Nearest Neighbor (ENN) rule; ADOMS, Adjusting the Direction Of the synthetic Minority clasS method; B-SMOTE, Borderline-SMOTE; DOS, Dynamic Over-Sampling; RUS, Random Under-Sampling; ADASYN, Adaptive Synthetic Sampling; SPIDER 1 and 2, frameworks that integrate a selective data pre-processing with an ensemble method; NCL, Neighborhood Cleaning Rule; TL, Tomek links method; STANDARD, back-propagation without any pre-processing; OSS, One-Sided Selection method; CNN, Condensed Nearest Neighbor; CNNTL, Condensed Nearest Neighbor with TL (for more details see Sections 3 and 4).

DATA	ROS	SDSAO	DyS	SDSAS	SMOTE-TL	SMOTE	SL-SMOTE	SMOTE-ENN	ADOMS	B-SMOTE	DOS	RUS	ADASYN	SPIDER-1	SPIDER-2	NCL	TL	STANDARD	OSS	CNN	CNNTL
CAY0	0.976	0.986	0.985	0.978	0.976	0.976	0.976	0.976	0.977	0.968	0.984	0.970	0.967	0.937	0.938	0.937	0.936	0.933	0.803	0.833	0.795
CAY1	0.969	0.979	0.975	0.968	0.969	0.969	0.970	0.970	0.970	0.966	0.985	0.955	0.965	0.735	0.789	0.745	0.717	0.752	0.931	0.916	0.918
CAY2	0.968	0.985	0.958	0.968	0.969	0.968	0.968	0.969	0.968	0.968	0.991	0.950	0.975	0.948	0.957	0.952	0.929	0.949	0.961	0.961	0.958
CAY3	0.985	0.985	0.985	0.985	0.986	0.985	0.984	0.985	0.984	0.973	0.991	0.950	0.967	0.948	0.949	0.940	0.929	0.941	0.809	0.826	0.839
CAY4	0.974	0.994	0.991	0.971	0.971	0.968	0.971	0.968	0.968	0.964	0.962	0.922	0.967	0.914	0.936	0.888	0.865	0.846	0.946	0.934	0.956
CAY5	0.952	0.922	0.956	0.943	0.952	0.950	0.949	0.951	0.951	0.950	0.908	0.933	0.951	0.781	0.834	0.773	0.769	0.772	0.666	0.618	0.617
CAY6	0.980	0.956	0.956	0.980	0.981	0.981	0.980	0.982	0.980	0.969	0.956	0.960	0.973	0.946	0.946	0.952	0.949	0.830	0.850	0.787	0.875
CAY7	0.991	0.983	0.983	0.990	0.990	0.990	0.991	0.991	0.991	0.983	0.986	0.986	0.967	0.986	0.985	0.984	0.984	0.985	0.776	0.824	0.788
CAY8	0.975	0.937	0.935	0.966	0.976	0.972	0.971	0.972	0.970	0.964	0.923	0.925	0.958	0.933	0.933	0.935	0.934	0.935	0.817	0.826	0.825
CAY9	0.915	0.923	0.920	0.910	0.917	0.916	0.915	0.915	0.916	0.911	0.898	0.896	0.909	0.875	0.868	0.860	0.848	0.834	0.879	0.849	0.872
CAY10	0.967	0.979	0.965	0.968	0.963	0.968	0.968	0.969	0.968	0.965	0.973	0.885	0.970	0.883	0.902	0.860	0.877	0.922	0.833	0.786	0.808
FELT0	0.979	0.982	0.982	0.979	0.978	0.978	0.977	0.977	0.978	0.971	0.977	0.951	0.951	0.976	0.976	0.977	0.976	0.977	0.952	0.955	0.937
FELT1	0.976	0.966	0.98	0.975	0.976	0.973	0.975	0.976	0.976	0.968	0.971	0.970	0.976	0.964	0.964	0.965	0.964	0.952	0.947	0.946	0.945
FELT2	0.976	0.947	0.960	0.976	0.974	0.975	0.975	0.974	0.975	0.959	0.969	0.959	0.963	0.914	0.921	0.918	0.901	0.890	0.952	0.948	0.948
FELT3	0.977	0.984	0.987	0.978	0.977	0.978	0.977	0.977	0.978	0.968	0.987	0.971	0.956	0.974	0.975	0.969	0.971	0.970	0.964	0.966	0.962
FELT4	0.983	0.992	0.976	0.985	0.983	0.983	0.984	0.983	0.983	0.977	0.988	0.981	0.964	0.968	0.972	0.972	0.968	0.968	0.968	0.969	0.961
SAT0	0.920	0.910	0.916	0.917	0.915	0.915	0.916	0.914	0.916	0.918	1.000	0.913	0.907	0.909	0.912	0.909	0.894	0.881	0.899	0.881	0.865
SAT1	0.985	0.988	0.988	0.984	0.986	0.983	0.986	0.984	0.985	0.983	0.996	0.983	0.976	0.983	0.983	0.982	0.982	0.981	0.982	0.981	0.976
SAT2	0.981	0.989	0.983	0.980	0.982	0.980	0.981	0.983	0.980	0.977	0.961	0.980	0.969	0.977	0.980	0.977	0.976	0.976	0.971	0.966	0.964
SAT3	0.961	0.965	0.958	0.940	0.958	0.962	0.962	0.961	0.963	0.960	0.911	0.957	0.955	0.957	0.958	0.950	0.955	0.943	0.954	0.956	0.945
SAT4	0.857	0.867	0.803	0.863	0.866	0.849	0.858	0.858	0.858	0.844	1.000	0.844	0.854	0.746	0.779	0.792	0.757	0.581	0.769	0.711	0.776
SAT5	0.944	0.945	0.928	0.925	0.944	0.944	0.944	0.942	0.941	0.920	1.000	0.917	0.913	0.847	0.823	0.855	0.853	0.842	0.827	0.921	0.927
SEG0	0.998	0.965	0.970	0.995	0.993	0.994	0.996	0.992	0.988	0.997	0.895	0.995	0.993	0.993	0.992	0.994	0.993	0.994	0.925	0.993	0.994
SEG1	1.000	1.000	1.000	1.000	1.000	1.000	1.000	1.000	0.988	1.000	0.991	0.999	0.991	1.000	1.000	0.999	0.999	1.000	0.999	0.909	0.914
SEG2	0.978	0.979	0.996	0.979	0.975	0.977	0.979	0.974	0.977	0.981	0.983	0.978	0.965	0.982	0.983	0.981	0.980	0.977	0.965	0.964	0.963
SEG3	0.973	0.967	0.976	0.963	0.971	0.973	0.972	0.970	0.911	0.971	0.952	0.961	0.961	0.969	0.966	0.970	0.974	0.957	0.961	0.960	0.957
SEG4	0.872	0.836	0.907	0.926	0.927	0.906	0.852	0.926	0.514	0.863	0.850	0.886	0.903	0.787	0.811	0.650	0.656	0.630	0.793	0.741	0.840
SEG5	0.999	1.000	1.000	0.999	0.994	0.993	0.981	0.992	0.980	0.998	0.957	0.995	0.990	0.995	0.998	0.994	0.990	0.999	0.918	0.996	0.994
SEG6	0.995	1.000	1.000	0.995	0.995	0.995	0.995	0.995	0.994	0.995	0.970	0.995	0.987	0.995	0.995	0.995	0.995	0.995	0.912	0.922	0.887
92A0	0.937	0.963	0.940	0.945	0.947	0.942	0.921	0.937	0.943	0.927	0.862	0.928	0.939	0.926	0.918	0.921	0.926	0.845	0.924	0.894	0.922
92A1	0.881	0.902	0.942	0.910	0.910	0.896	0.825	0.910	0.918	0.867	0.948	0.908	0.899	0.854	0.865	0.862	0.858	0.704	0.277	0.767	0.868
92A2	0.853	0.861	0.858	0.861	0.840	0.848	0.845	0.852	0.856	0.881	0.833	0.850	0.842	0.848	0.828	0.843	0.843	0.838	0.774	0.846	0.786
92A3	0.880	0.869	0.858	0.874	0.840	0.879	0.880	0.852	0.882	0.881	0.867	0.879	0.877	0.860	0.802	0.817	0.849	0.876	0.774	0.829	0.683
92A4	0.987	0.997	0.997	0.981	0.975	0.980	0.977	0.974	0.982	0.986	0.983	0.974	0.973	0.977	0.974	0.975	0.976	0.968	0.977	0.975	0.975
92A5	0.995	1.000	1.000	0.993	0.987	0.978	0.965	0.985	0.989	0.990	1.000	0.974	0.977	0.988	0.987	0.968	0.955	0.971	0.946	0.902	0.912
Ranks	5.500	6.129	6.171	6.471	7.000	7.143	7.443	7.643	7.857	9.600	9.871	11.700	12.971	13.586	13.771	14.900	15.586	16.029	16.871	17.229	17.629

In accordance with Table 2 and Figure 1, most methods of over-sampling present a better classification performance than the standard BP with statistical significance. The under-sampling methods do not present a statistical difference with respect to standard BP performance, and all dynamic sampling approaches improve the standard BP performance with statistical differences.

ADASYIN, SPIDER-1 and SPIDER-2 (over-sampling methods) and RUS, NCL and TL (under-sampling methods) show the trend of improving the classification results, but they do not significantly improve the standard BP performance. Then, the OSS, CNN and CNNTL classify worse than standard BP; this notwithstanding, these approaches do not show a statistical difference with it.

SDSAO, SDSAS and DyS are statistically better than ADASYIN, SPIDER-1 and SPIDER-2 (over-sampling methods) and also than all under-sampling approaches studied in this work. With a statistical difference, the DOS performance is better than CNN, CNNTL, OSS and TL.

Table 2 shows that the trend is that ROS presents a better performance than the proposed method (SDSAO and SDSAS), and that DyS shows a slight advantage over SDSAS; however, in accordance with the Holm–Shaffer post-hoc tests, statistical difference in the classification performance does not exist among these methods (see Figure 1).

In general terms, most over-sampling methods and dynamic sampling approaches are successful methods to deal with the class imbalance problem, but with respect to the training dataset size, SDSAS, SDSAO and DyS use significantly fewer samples than the over-sampling approaches. They employed about 78% less samples than most over-sampling methods; in addition, SDSAS, SDSAO and DyS still use fewer samples than the standard BP trained with the original training dataset. They use about 60% less samples; these facts stand out in Figure 2. However, the DyS method applies the ROS in each epoch or iteration (see Section 4), whereas SDSA only applies the ROS or SMOTE one time before ANN training (see Section 2).

Figure 2 shows that the under-sampling methods employ significantly fewer samples than the rest of the techniques (except dynamic sampling approaches with respect to RUS, NCL and TL); however, their classification performance in most of the cases is worse than the standard BP (without statistical significant) or is not better (with statistical significant) than the standard BP.

On the other hand, the worst methods studied in this paper (in agreement with Table 2) are those based on the CNN technique (OSS, CNN and CNNTL), i.e., those that use a $k - NN$ rule as the basis and achieving an important size reduction of the training dataset. In contrast, NCL, which is of the $k - NN$ family, also improves the classification performance of the back-propagation; however, the dataset size reduction reached for this method is not of CNN's magnitude; in addition, it only eliminates majority samples. The use of TL (TL and SMOTE-TL) seems to increase the classification performance, but it does not eliminate too many samples (see Figure 2), except by CNNTL, which we consider to cancel the positive effect of TL by the important training dataset reduction. SMOTE-ENN does not seem to improve the classification performance of SMOTE in spite of including a cleaning step that removes both majority and minority samples. The methods that have achieved the enhancing of the classifier performance are those that only eliminate samples from the majority class.

Furthermore, analyzing only the selective samples methods (SL-SMOTE, B-SMOTE, ADASYN, SPIDER-1 and SPIDER-2), those are the ones in which the more appropriate samples are selected to be over-sampled. It is considered that in the result presented in Figure 2, SL-SMOTE and B-SMOTE obtain the best results, whereas the advantages of ADASYN, SPIDER-1 and SPIDER-2 are not clear (RUS often outperforms these approaches, but without statistical significance; Figure 1). SL-SMOTE, B-SMOTE and the proposed method do not show statistical significance in their classification results, but the number of samples used by SDSA in the training stage is fewer than employed for SL-SMOTE and B-SMOTE (see Figure 2).

Focusing on the dynamic sampling approaches' analysis, SDSAO presents a slight advantage in performance than DyS and SDSAS, whereas DOS does not seem to be an attractive method. However, the aim of DOS is to identify a suitable over-sampling rate, whilst reducing the processing time and storage requirements, as well as keeping or increasing the ANN performance, to obtain a trade-off between classification performance and computational cost.

SDSA and DyS improve the classification performance, including a selective process, but while DyS tries to reduce the oversampling ratio during the training (i.e., it applies the ROS method in each epoch with different class imbalance ratios; see Section 4), the SDSA only tries to use the "best samples" to train the ANN.

Dynamic sampling approaches are a very attractive way to deal with a class imbalance problem. They face two important topics: (i) improving the classification performance; and (ii) reducing the classifier computational cost.

7. Conclusions and Future Work

We propose a new Selective Dynamic Sampling Approach (SDSA) to deal with the class imbalance problem. It is attractive because it automatically selects the best samples to train the multilayer perceptron neural network with the stochastic back-propagation. The SDSA identifies the most appropriate samples ("average samples") to train the neural network. The average samples are the most adequate samples to train the neural network; they are neither hard nor easy to learn. These are between the safe and border areas in the training space. SDSA employs a Gaussian function to give priority to the average samples during the neural network training stage.

The experimental results in this paper point out that SDSA is a successful method to deal with the class imbalance problem, and its performance is statistically equivalent to other well-known over-sampling and dynamic sampling approaches. It is statistically better than the under-sampling methods compared to this work and also than the standard back-propagation. In addition, in the neural network training stage, SDSA uses significantly fewer samples than the over-sampling methods, even than the standard back-propagation trained with the original dataset.

Future work will extend this study. The interest is: to explore the effectiveness of the SDSA in multi-class and high imbalanced problems and to find a mechanism to automatically identify the most suitable μ value for each dataset. The appropriate selection of μ value might significantly improve the proposed method. In addition, it is important to explore the possibility to use the SDSA to obtain optimal subsets to train other classifiers like support vector machines or to compare its effectiveness with the other kinds of class imbalance approaches using other learning models.

Acknowledgments: This work has partially been supported by Tecnológico de Estudios Superiores de Jocotitlán under grant SDMAIA-010.

Author Contributions: Roberto Alejo, Juan Monroy-de-Jesús and Juan H. Pacheco-Sánchez conceived and designed the experiments. Erika López-González performed the experiments. Juan A. Antonio-Velázquez analyzed the data. Roberto Alejo and Juan H. Pacheco-Sánchez wrote the paper.

Conflicts of Interest: The authors declare no conflict of interest.

References

1. Prati, R.C.; Batista, G.E.A.P.A.; Monard, M.C. Data mining with imbalanced class distributions: Concepts and methods. In Proceedings of the 4th Indian International Conference on Artificial Intelligence (IICAI 2009), Tumkur, Karnataka, India, 16–18 December 2009; pp. 359–376.
2. Galar, M.; Fernández, A.; Tartas, E.B.; Sola, H.B.; Herrera, F. A review on ensembles for the class imbalance problem: Bagging-, boosting-, and hybrid-based approaches. *IEEE Trans. Syst. Man Cyber. Part C* **2012**, *42*, 463–484.
3. García, V.; Sánchez, J.S.; Mollineda, R.A.; Alejo, R.; Sotoca, J.M. The class imbalance problem in pattern classication and learning. In *II Congreso Español de Informática*; Thomson: Zaragoza, Spain, 2007; pp. 283–291.
4. Wang, S.; Yao, X. Multiclass imbalance problems: Analysis and potential solutions. *IEEE Trans. Syst. Man Cyber. Part B* **2012**, *42*, 1119–1130.
5. Nanni, L.; Fantozzi, C.; Lazzarini, N. Coupling different methods for overcoming the class imbalance problem. *Neurocomput* **2015**, *158*, 48–61.
6. Loyola-González, O.; Martínez-Trinidad, J.F.; Carrasco-Ochoa, J.A.; García-Borroto, M. Study of the impact of resampling methods for contrast pattern based classifiers in imbalanced databases. *Neurocomputing* **2016**, *175*, 935–947.

7. López, V.; Fernández, A.; García, S.; Palade, V.; Herrera, F. An insight into classification with imbalanced data: Empirical results and current trends on using data intrinsic characteristics. *Inf. Sci.* **2013**, *250*, 113–141.

8. Chawla, N.V.; Bowyer, K.W.; Hall, L.O.; Kegelmeyer, W.P. SMOTE: Synthetic minority over-sampling technique. *J. Artif. Intell. Res.* **2002**, *16*, 321–357.

9. Batista, G.E.A.P.A.; Prati, R.C.; Monard, M.C. A study of the behavior of several methods for balancing machine learning training data. *SIGKDD Explor. Newsl.* **2004**, *6*, 20–29.

10. He, H.; Garcia, E. Learning from imbalanced data. *IEEE Trans. Knowl. Data Eng.* **2009**, *21*, 1263–1284.

11. Han, H.; Wang, W.; Mao, B. Borderline-SMOTE: A New Over-Sampling Method in Imbalanced Data Sets Learning. In Proceedings of the International Conference on Intelligent Computing (ICIC 2005), Hefei, China, 23–26 August 2005; pp. 878–887.

12. Bunkhumpornpat, C.; Sinapiromsaran, K.; Lursinsap, C. Safe-level-SMOTE: Safe-level-synthetic minority over-sampling technique for handling the class imbalanced problem. In Proceedings of the 13th Pacific-Asia Conference (PAKDD 2009), Bangkok, Thailand, 27–30 April 2009; Volume 5476, pp. 475–482.

13. He, H.; Bai, Y.; Garcia, E.A.; Li, S. ADASYN: Adaptive Synthetic Sampling Approach for Imbalanced Learning. In Proceedings of the 2008 IEEE International Joint Conference on Neural Networks (IJCNN 2008), Hong Kong, China, 1–8 June 2008; pp. 1322–1328.

14. Stefanowski, J.; Wilk, S. Selective pre-processing of imbalanced data for improving classification performance. In Proceedings of the 10th International Conference in Data Warehousing and Knowledge Discovery (DaWaK 2008), Turin, Italy, 1–5 September 2008; pp. 283–292.

15. Napierala, K.; Stefanowski, J.; Wilk, S. Learning from Imbalanced Data in Presence of Noisy and Borderline Examples. In Proceedings of the 7th International Conference on Rough Sets and Current Trends in Computing (RSCTC 2010), Warsaw, Poland, 28–30 June 2010; pp. 158–167.

16. Japkowicz, N.; Stephen, S. The class imbalance problem: A systematic study. *Intell. Data Anal.* **2002**, *6*, 429–449.

17. Liu, X.; Wu, J.; Zhou, Z. Exploratory undersampling for class-imbalance learning. *IEEE Trans. Syst. Man Cyber. Part B* **2009**, *39*, 539–550.

18. Alejo, R.; Valdovinos, R.M.; García, V.; Pacheco-Sanchez, J.H. A hybrid method to face class overlap and class imbalance on neural networks and multi-class scenarios. *Pattern Recognit. Lett.* **2013**, *34*, 380–388.

19. Alejo, R.; García, V.; Pacheco-Sánchez, J.H. An efficient over-sampling approach based on mean square error back-propagation for dealing with the multi-class imbalance problem. *Neural Process. Lett.* **2015**, *42*, 603–617.

20. Wilson, D. Asymptotic properties of nearest neighbor rules using edited data. *IEEE Trans. Syst. Man Cyber.* **1972**, *2*, 408–420.

21. Kubat, M.; Matwin, S. Addressing the Curse of Imbalanced Training Sets: One-sided Selection. In Proceedings of the 14th International Conference on Machine Learning (ICML 1997), Nashville, TN, USA, 8–12 July, 1997; pp. 179–186.

22. Tomek, I. Two modifications of CNN. *IEEE Trans. Syst. Man Cyber.* **1976**, *7*, 679–772.

23. Hart, P. The condensed nearest neighbour rule. *IEEE Trans. Inf. Theory* **1968**, *14*, 515–516.

24. Zhou, Z.H.; Liu, X.Y. Training cost-sensitive neural networks with methods addressing the class imbalance problem. *IEEE Trans. Knowl. Data Eng.* **2006**, *18*, 63–77.

25. Drummond, C.; Holte, R.C. Class Imbalance, and Cost Sensitivity: Why Under-Sampling Beats Over-Sampling. In Proceedings of the Workshop on Learning from Imbalanced Datasets II, (ICML 2003), Washington, DC, USA, 2003; pp. 1–8. Available online: http://citeseerx.ist.psu.edu/viewdoc/summary?doi=10.1.1.132.9672 (accessed on 4 July 2016).

26. He, H.; Ma, Y. *Imbalanced Learning: Foundations, Algorithms, and Applications*, 1st ed.; John Wiley & Sons, Inc Press: Hoboken, NJ, USA, 2013.

27. Lin, M.; Tang, K.; Yao, X. Dynamic sampling approach to training neural networks for multiclass imbalance classification. *IEEE Trans. Neural Netw. Learn. Syst.* **2013**, *24*, 647–660.

28. Wang, J.; Jean, J. Resolving multifont character confusion with neural networks. *Pattern Recognit.* **1993**, *26*, 175–187.

29. Ou, G.; Murphey, Y.L. Multi-class pattern classification using neural networks. *Pattern Recognit.* **2007**, *40*, 4–18.

30. Murphey, Y.L.; Guo, H.; Feldkamp, L.A. Neural learning from unbalanced data. *Appl. Intell.* **2004**, *21*, 117–128.

31. Fernández-Navarro, F.; Hervás-Martínez, C.; Antonio Gutiérrez, P. A dynamic over-sampling procedure based on sensitivity for multi-class problems. *Pattern Recognit.* **2011**, *44*, 1821–1833.

32. Fernández-Navarro, F.; Hervás-Martínez, C.; García-Alonso, C.R.; Torres-Jiménez, M. Determination of relative agrarian technical efficiency by a dynamic over-sampling procedure guided by minimum sensitivity. *Expert Syst. Appl.* **2011**, *38*, 12483–12490.

33. Chawla, N.V.; Cieslak, D.A.; Hall, L.O.; Joshi, A. Automatically countering imbalance and its empirical relationship to cost. *Data Min. Knowl. Discov.* **2008**, *17*, 225–252.

34. Debowski, B.; Areibi, S.; Gréwal, G.; Tempelman, J. A Dynamic Sampling Framework for Multi-Class Imbalanced Data. In Proceedings of the Machine Learning and Applications (ICMLA), 2012 11th International Conference on (ICMLA 2012), Boca Raton, FL, USA, 12–15 December 2012; pp. 113–118.

35. Alejo, R.; Monroy-de-Jesus, J.; Pacheco-Sanchez, J.; Valdovinos, R.; Antonio-Velazquez, J.; Marcial-Romero, J. Analysing the Safe, Average and Border Samples on Two-Class Imbalance Problems in the Back-Propagation Domain. In *Progress in Pattern Recognition, Image Analysis, Computer Vision, and Applications—CIARP 2015*; Springer-Verlag: Montevideo, Uruguay, 2015; pp. 699–707.

36. Lawrence, S.; Burns, I.; Back, A.; Tsoi, A.; Giles, C.L. Neural network classification and unequal prior class probabilities. *Neural Netw. Tricks Trade* **1998**, *1524*, 299–314.

37. Laurikkala, J. Improving Identification of Difficult Small Classes by Balancing Class Distribution. In Proceedings of the 8th Conference on AI in Medicine in Europe: Artificial Intelligence Medicine (AIME 2001), Cascais, Portugal, 1–4 July 2001; pp. 63–66.

38. Prati, R.; Batista, G.; Monard, M. Class Imbalances versus Class Overlapping: An Analysis of a Learning System Behavior. In Proceedings of the Third Mexican International Conference on Artificial Intelligence (MICAI 2004), Mexico City, Mexico, 26–30 April 2004; pp. 312–321.

39. Batista, G.E.A.P.A.; Prati, R.C.; Monard, M.C. Balancing Strategies and Class Overlapping; In Proceedings of the 6th International Symposium on Intelligent Data Analysis, (IDA 2005), Madrid, Spain, 8–10 September 2005; pp. 24–35.

40. Tang, Y.; Gao, J. Improved classification for problem involving overlapping patterns. *IEICE Trans.* **2007**, *90-D*, 1787–1795.

41. Inderjeet, M.; Zhang, I. KNN approach to unbalanced data distributions: A case study involving information extraction. In Proceedings of the International Conference on Machine Learning (ICML 2003), Workshop on Learning from Imbalanced Data Sets, Washington, DC, USA, 21 August 2003.

42. Stefanowski, J. Overlapping, Rare Examples and Class Decomposition in Learning Classifiers from Imbalanced Data. In *Emerging Paradigms in Machine Learning*; Ramanna, S., Jain, L.C., Howlett, R.J., Eds.; Springer Berlin Heidelberg: Berlin, Germany, 2013; Volume 13, pp. 277–306.

43. Duda, R.; Hart, P.; Stork, D. *Pattern Classification*, 2nd ed.; Wiley: New York, NY, USA, 2001.

44. Fawcett, T. An introduction to ROC analysis. *Pattern Recogn. Lett.* **2006**, *27*, 861–874.

45. Tang, S.; Chen, S. The Generation Mechanism of Synthetic Minority Class Examples. In Proceedings of the 5th International Conference on Information Technology and Applications in Biomedicine (ITAB 2008), Shenzhen, China, 30–31 May 2008; pp. 444–447.

46. Bruzzone, L.; Serpico, S. Classification of imbalanced remote-sensing data by neural networks. *Pattern Recognit. Lett.* **1997**, *18*, 1323–1328.

47. Lichman, M. *UCI Machine Learning Repository*; University of California, Irvine, School of Information and Computer Sciences: Irvine, CA, USA, 2013.

48. Baumgardner, M.; Biehl, L.; Landgrebe, D. *220 Band AVIRIS Hyperspectral Image Data Set: June 12, 1992 Indian Pine Test Site 3*; Purdue University Research Repository, School of Electrical and Computer Engineering, ITaP and LARS: West Lafayette, IN, USA, 2015.

49. Madisch, I.; Hofmayer, S.; Fickenscher, H. *Roberto Alejo*; ResearchGate: San Francisco, CA, USA, 2016.

50. Wilson, D.R.; Martinez, T.R. Improved heterogeneous distance functions. *J. Artif. Int. Res.* **1997**, *6*, 1–34.

51. Alcalá-Fdez, J.; Fernandez, A.; Luengo, J.; Derrac, J.; García, S.; Sánchez, L.; Herrera, F. KEEL data-mining software tool: Data set repository, integration of algorithms and experimental analysis framework. *J. Mult. Valued Logic Soft Comput.* **2011**, *17*, 255–287.

52. Iman, R.L.; Davenport, J.M. Approximations of the critical region of the friedman statistic. *Commun. Stat. Theory Methods* **1980**, *9*, 571–595.

53. Holm, S. A simple sequentially rejective multiple test procedure. *Scand. J. Stat.* **1979**, *6*, 65–70.

54. Luengo, J.; García, S.; Herrera, F. A study on the use of statistical tests for experimentation with neural networks: Analysis of parametric test conditions and non-parametric tests. *Expert Syst. Appl.* **2009**, *36*, 7798–7808.

55. García, S.; Herrera, F. An extension on "statistical comparisons of classifiers over multiple data sets" for all pairwise comparisons. *J. Mach. Learn. Res.* **2008**, *9*, 2677–2694.

56. García, V.; Mollineda, R.A.; Sánchez, J.S. On the k-NN performance in a challenging scenario of imbalance and overlapping. *Pattern Anal. Appl.* **2008**, *11*, 269–280.

57. Denil, M.; Trappenberg, T.P. Overlap versus Imbalance. In Proceedings of the Canadian Conference on AI, Ottawa, NO, Canada, 31 May–2 June 2010; pp. 220–231.

58. Jo, T.; Japkowicz, N. Class imbalances versus small disjuncts. *SIGKDD Explor. Newsl.* **2004**, *6*, 40–49.

59. Ertekin, S.; Huang, J.; Bottou, L.; Giles, C. Learning on the border: Active learning in imbalanced data classification. In Proceedings of the ACM Conference on Information and Knowledge Management, Lisbon, Portugal, 6–10 November 2007.

NHL and RCGA Based Multi-Relational Fuzzy Cognitive Map Modeling for Complex Systems

Zhen Peng, Lifeng Wu and Zhenguo Chen

Abstract: In order to model multi-dimensions and multi-granularities oriented complex systems, this paper firstly proposes a kind of multi-relational Fuzzy Cognitive Map (FCM) to simulate the multi-relational system and its auto construct algorithm integrating Nonlinear Hebbian Learning (NHL) and Real Code Genetic Algorithm (RCGA). The multi-relational FCM fits to model the complex system with multi-dimensions and multi-granularities. The auto construct algorithm can learn the multi-relational FCM from multi-relational data resources to eliminate human intervention. The Multi-Relational Data Mining (MRDM) algorithm integrates multi-instance oriented NHL and RCGA of FCM. NHL is extended to mine the causal relationships between coarse-granularity concept and its fined-granularity concepts driven by multi-instances in the multi-relational system. RCGA is used to establish high-quality high-level FCM driven by data. The multi-relational FCM and the integrating algorithm have been applied in complex system of Mutagenesis. The experiment demonstrates not only that they get better classification accuracy, but it also shows the causal relationships among the concepts of the system.

Reprinted from *Appl. Sci.*. Cite as: Peng, Z.; Wu, L.; Chen, Z. NHL and RCGA Based Multi-Relational Fuzzy Cognitive Map Modeling for Complex Systems. *Appl. Sci.* **2015**, *5*, 1399–1411.

1. Introduction

The aim of the paper is to auto simulate complex systems with multi-dimensions and multi-granularities driven by multi-relational data resources for better classification and causal relationships of a system.

Multi-Relational Data Mining (MRDM) [1,2] is able to discover knowledge directly from multi-relational data tables, not through connection and aggregation of multiple relational data into a single data. Multi-relationship data mining can effectively prevent the problems of information loss, statistical deviation and low efficiency, *etc.* These methods [3–5] such as CrossMine, MI-MRNBC and Graph-NB are fitting for multi-relational data mining, but cannot obtain causality in a multi-relational system.

In 1986, FCM [6–8] is introduced by Kosko, suggesting the use of fuzzy causal functions taking numbers in [–1, 1] in concept maps. FCM, as a kind of graph model, combines some aspects from fuzzy logic, neural networks and other techniques, and is fitting for modeling system from data resources. Compared to other techniques, FCM exhibits a number of highly appealing properties. In particular, FCM can directly show the multi-relationships between different concepts and the inference is easy and intuitive. FCM learning algorithms use learning algorithms to establish models from historical data (simulations of concept values), which have been widely used applied to various fields [9–13] of society, engineering, medicine, environmental science, *etc.* However, none of them consider the data characteristics of multi-relationships.

Multi-relational FCM discussed in the paper refers to two-levels FCM. There is one FCM of each dimension in low-level. There is only one FCM in high-level. The state value of each concept in high-level actually is a summary evaluation of low-level FCM in the dimension, which is inferred based on the weight vector, obtained by multi-instances oriented NHL, in low-level FCM. RCGA of high-level FCM aims to mine high-level FCM based on summary evaluations of low-level FCMs for high-quality classification and causality. Thus, the proposed multi-relational FCM and the integrating algorithm seem a rather realistic approach to solve the complex model.

The remainder of this paper is organized as follows. Section 2 describes FCM, existing learning algorithms of FCM, and problems to solve. Then, Section 3 proposes the multi-relational FCM and integrating NHL and RCGA based multi-relational FCM learning algorithm. In Section 4, the experiment and its results are represented and analyzed. Finally, we briefly conclude this paper in Section 5.

2. Backgrounds

2.1. Fuzzy Cognitive Map (FCM)

A Fuzzy Cognitive Map F in Figure 1 shows a relationship system, which is a 4-tuple (C, W, A, f) mathematically, where

- $C = \{C_1, C_2, \ldots, C_N\}$ is a set of N concepts forming the nodes of a graph.
- $W: (C_i, C_j) \rightarrow w_{ij}$ is a function associating w_{ij} with a pair of concepts, with w_{ij} equal to the weight of edge directed from C_i to C_j, where $w_{ij} \in [-1, 1]$. Thus, $W(NN)$ is a connection matrix.

If there is positive causality between concepts C_i and C_j, then $w_{ij} > 0$, which means an increase of the value of concept C_i will cause an increase of the value of concept C_j and a decrease of the value of C_i will lead to a decrease of the value of C_j.

If there is inverse causality between the two concepts, then $w_{ij} < 0$, which represents an increase of the value of concept C_i will cause a decrease of the value of the second concept and a decrease of the value of concept C_i will cause an increase of the value of the concept C_j.

If there is no relationship between the two concepts, then $w_{ij} = 0$.

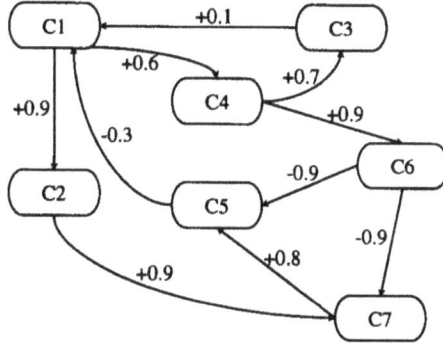

Figure 1. A simple fuzzy cognitive map.

- $A: C_i \rightarrow A_i(t)$ is a function that associates each concept C_i with the sequence of its activation degrees such as for $t \epsilon T$, $A_i(t) \epsilon L$ given its activation degree at the moment t. $A(0) \epsilon L^T$ indicates the initial vector and specifies initial values of all concept nodes and $A(t) \epsilon L^T$ is a state vector at certain iteration t.
- f is a transformation or activation function, which includes recurring relationship on $t \geq 0$ between $A(t + 1)$ and $A(t)$.

$$A_j (t+1) = f(A_j (t) + \sum_{\substack{i \neq j \\ i \in S}} A_i (t) w_{ij}) \tag{1}$$

The transformation function of FCM is with memory at previous moment shown as Equation (1), which is used to infer the state values of concepts in FCM. It limits the weighted sum to the range [0, 1] or [−1, 1]. The three most commonly used transformation (bivalent, trivalent, logistic) functions are shown below.

i   Bivalent

$$f(x) = \begin{cases} 0, x \leq 0 \\ 1, \ x > 0 \end{cases}$$

ii   Trivalent

$$f(x) = \begin{cases} -1, \ x \leq -0.5 \\ 0, \ -0.5 \leq x \leq 0.5 \\ 1, \ x \geq 0.5 \end{cases}$$

iii   Logistic

$$f(x) = \frac{1}{1 + e^{-Cx}}$$

FCM can be used to perform simulation of an interconnected system. The vector $A(t)$ in FCM specifies state values of all concepts (nodes) in the t iteration. An FCM has a number of successive state iterations. The state value of a concept is calculated by the preceding iteration of concepts states, which exert influence on the given node.

2.2. FCM Learning Algorithms

FCM learning algorithm is a kind of automated learning method to establish FCM model from data resources. There are two classes of FCM learning algorithms, Hebbian based learning and evolved based learning. The former are Hebbian based algorithms [14–16], mainly including NHL (Nonlinear Hebbian Learning), DD-NHL (Data-Driven Nonlinear Hebbian Learning) and AHL (Active Hebbian Learning). The differences of these algorithms are in the way of adjusting the edge weights. The latter are evolve based algorithms [17–20], which are composed of PSO (Particle Swarm Optimization), RCGA (Real Coded Genetic Algorithm), *etc.*

2.2.1. Nonlinear Hebbian Learning (NHL)

NHL is on the basis of the well-known Hebb's learning law, which is a kind of unsupervised learning algorithm. Considering the nonlinear output unit, given random pre-synaptic an input vector x, weight vector w, and output $z = f\ (w^T\ x)$. The nonlinear activation function f is a sigmoid function. The criterion function J maximized by Hebb's rule may be written as Equation (2).

$$J = E[z^2] \tag{2}$$

An additional constraint such as $||w|| = 1$ is necessary to stabilize the learning rule. A stochastic approximation solution is employed to the following nonlinear Hebbian learning rule as Equation (3).

$$\Delta w_{ji} = \eta_k z \frac{dz}{dy}(x_j - w_{ji}y_i) \tag{3}$$

Note that the nonlinear learning rules are seeking a set of weight parameters such that the outputs of the unit have the largest variance. The nonlinear unit constrains the output, ensuring it remains within a bounded range.

2.2.2. Real-Coded Genetic Algorithm (RCGA)

RCGA is a real-coded genetic algorithm to develop FCM connection matrix based on data resource. RCGA defines each chromosome as a floating-point vector. Each element in the vector is called gene. In case of the learning FCM with N node, each chromosome consists of $N(N-1)$ genes, which are floating point numbers from the range $[-1, 1]$, defined as follows.

$$E = [w_{12}, w_{13}, \ldots, w_{1N}, w_{21}, w_{23}, \ldots, w_{2N}, \ldots, w_{NN-1}]^T \qquad (4)$$

where w_{ij} is a weight value for an edge from i^{th} to j^{th} concept node. Each chromosome has to be decoded back into a candidate FCM. The number of chromosomes in a population is constant for each generation and it is specified by the *population_size* parameter.

The fitness function is calculated for each chromosome by computing the difference between system response generated using a FCM weights and a corresponding system response, which is known directly from the data resource. The difference is computed across all $M-1$ initial vector/system response pairs, and for the same initial state vector. The measure of error is shown as Equation (5).

$$Error\_Lp = \alpha \sum_{t=1}^{M-1} \sum_{n=1}^{N} \left| A_n(t) - \widehat{A_n(t)} \right|^p \qquad (5)$$

The parameter of α is used to normalize error rate, and p is $1, 2$ or ∞. N is the number of concepts in FCM, and M is the number of iterations. The error measure can be used as the core of fitness function as Equation (6).

$$\text{Fitness function} = \frac{1}{a \times Error\_Lp + 1} \qquad (6)$$

The fitness function is normalized to the $(0, 1]$. The parameter a can be set different value in different p condition.

The stopping condition of RCGA takes into consideration two possible scenarios of the learning process. One is the learning should be terminated when the fitness function value reaches a threshold value called *max_fitness*; the other is a maximum number of generations, named *max_generation*, has been reached. If the stopping conditions have not been reached, evolutionary operators and selection strategy need to be applied.

2.3. Problem Statements

In real world, a complex system has to have multi-dimensional groups with direct or indirect relationships, which generates multi-relational data. Moreover, each dimension maybe contains many concepts with different granularity relationships. For example, in a Mutagenesis system, there are three dimensions of atom, molecule and another atom, which are coarse-grained concepts; atype and charge are fine-grained concepts in the atom dimension; lumo, logp, indl and inda are fine-grained concepts in the molecule dimension. In two granularities, fine-grained concepts are on behalf of nodes in low-level and coarse-grained concepts represent hyper nodes. The coarse-grained concept of a dimension can be seen as a summary expression of its fine-grained concepts with multi-instances in low-level of the dimension.

Thus, problem to solve are: how to get summary evaluations in a dimension based on low-level data with multi-instances and how to mine high-level FCM from summary evaluations in dimensions for high-quality classification and causality.

3. Materials and Methods

3.1. Multi-Relational FCM

In order to better model the multi-relationships and coarse-grained concepts in the complex system, undoubtedly, a multi-relational FCM (Definition 1), extended from FCM, can represent the multi-relationship system. The multi-relational FCM can be divided into different groups and different levels. A group means a dimension. A level is a granularity. Coarse-grained concepts are upper-level nodes. Fine-grained concepts are low-level nodes. A coarse-grained concept in a dimension represents a FCM composed of fine-grained concepts, and is related with other coarse-grained concepts in other dimensions.

Definition 1. A multi-relational FCM with two-levels and n-dimensions is $U_n{}^2 = (C_n{}^2, W_n{}^2, A_n{}^2, f)$.

- $C_n{}^2$: $\{\{C_{1i}\} \ldots ,\{C_{ji}\}, \ldots \{C_{ni}\}\}$ is a set of concepts, $\{C_{ji}\}$ is on behalf of a coarse-grained concept in j dimension, and C_{ji} is i^{th} fine-grained concept in bottom-level of jth dimension.
- $W_n{}^2$: $\{\{W_j\}, \{W_{ij}\}\}$. $<\{C_{ji}\}> \rightarrow W_j$ is a function associating W_j among j^{th} dimension, W_j:$\{w_{ij}\}$; $(<\{C_{1i}\}> \ldots , <\{C_{ji}\}>, \ldots <\{C_{ni}\}>) \rightarrow \{W_{ij}\}$ is function associating between coarse-grained concepts.
- $A_n{}^2$: $C_{ji} \rightarrow A_{ji}(t)$, $\{C_{ji}\} \rightarrow A_j(t)$. $A_j(t)$ is a function f at iteration t.
- f is a transformation function, which includes recurring relationship on $t \geq 0$ among $A_j(t + 1)$, $A_{ji}(t)$ and $A_i(t)$, where $A_j(0)$ is referred out based on the weight vector W_j, got by multi-instances oriented NHL, in low-level FCM.

3.2. Multi-Instances Oriented NHL

In the multi-relational FCM, each concept represents a useful field name in data resource. In the multi-relationship, one field in main table corresponding to another table has some sub fields. The field in the main table points to a coarse-grained concept in low-level FCM and the fields in another table indicate the fine-grained concepts in the FCM. Thus, the low-level FCM can be used to model the multi-relationship.

One coarse-grained concept (C_j) in j^{th} dimension corresponds to some fine-grained concepts $(C_{j1}, C_{j2}, \ldots, C_{jn})$. It becomes the key to the state value of get the weights between the coarse-grained concept and some fine-grained concepts in the low-level FCM. The weights express the causality relationship between fine-gain concepts and coarse-grained concept. The weights need to be learned from multi-instances. NHL can be extended to multi-instances oriented mining for the optimistic of the nonlinear units' weights in Figure 2. In the prerequisite, the state states of coarse-grained concepts in high-level FCM can be inferred.

Multi-instances oriented NHL has two constraints. First constraint is that it maximizes the mathematical expectation of A_j^2 of all multi-instances as Equation (7).

$$\text{maxmize } J = E[\sum_r A_j^2] \tag{7}$$

$$\text{subject to } ||w|| = 1 \tag{8}$$

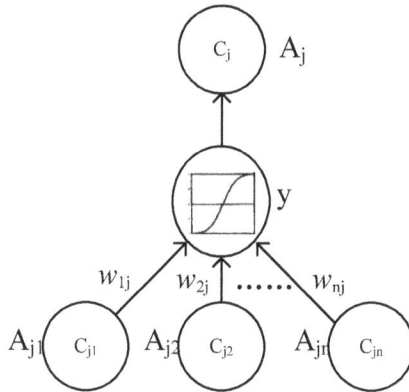

Figure 2. Nonlinear unit in Fuzzy Cognitive Map (FCM).

Second constraint is that the weight vector w has to be limited to stabilize the learning rule as Equation (8), which generates the following nonlinear Hebbian learning rule as Equation (9).

$$\Delta w_{ij}(t+1) = \eta A_j (A_{ji} - w_{ij}(t) A_j) \tag{9}$$

Accordingly, multi-instances oriented NHL is presented in the function of M_NHL. The execution phase of multi-instances oriented NHL (M_NHL) is consisted of the following steps:

Step 1: Random initialize the weight vector $w_j(t)$, $t = 0$, $p = 0$ and input all instance $\{A_{ji}\}$

Step 2: Calculate the mathematical expectation of A_j^2 of all $\{A_{ji}\}$

Step 3: Set $t = t + 1$, repeat for each iteration step t:

    3.1   Set $p = p + 1$, to the pth instance:

        3.1.1   Adjust $w_j(t)$ matrix to $\{A_{ji}\}^p$ by Equation (9)

    3.2   Calculate the A_j^2 to all $\{A_{ji}\}$ by Equation (10)

    3.3   Determine whether A_j^2 is maximum or not at present

    3.4   If A_j^2 is maximum, output the optimal $w_j(t)$

Step 4: Return the final weight vector $w_j(t)$

According to the literature [21], whether NHL clusters better or not is closely related to the activation function f. It means that if cumulative normal distribution function or approximate cumulative logic distribution function is chosen as activation function, output results will show a U-distribution and easily achieve better NHL clustering. In order to avoid integral calculation, the cumulative logic distribution, shown as Equation (11), is selected. The output values (nonlinear unit outputs) can be inferred by Equation (10), where A_{ij} is the input and w_{ji} is mined by M_NHL algorithm.

$$A_j(t) = f\left(\sum_{i \neq j} A_{ij}(t-1)w_{ji}(t-1)\right) \tag{10}$$

$$f(x) = \frac{1 - e^{-x}}{1 + e^{-x}} \tag{11}$$

3.3. NHL and RCGA Based Integrated Algorithm

RCGA is used to get the weight matrix of high-level FCM for high-quality classification and causality based on initial state values of coarse-grained concepts, which are got by M_NHL in Section 3.2.

$$W = [W_{12}, W_{13}, \ldots, W_{1M}, \ldots, W_{M1}, \ldots, W_{M(M-1)}] \tag{12}$$

Each chromosome consists of $M(M-1)$ genes (see Equation (12)). M is the total number of hyper concepts or dimensions. The gene is a floating point number from the range $[-1, 1]$. W_{ij} specifies the weight between coarse-grained concept in i^{th} dimension and it in j^{th} dimension. Each chromosome can be decoded back into a high-level FCM.

When the *fitness* is more than *max_fitness* specified or the generation is more than *max_generation* specified, the procedure ends. The fitness function of RCGA is as follows (see Equation (13)).

$$fitness = \frac{1}{a\Sigma_{t=0}^{T}\Sigma_{j=1}^{M}\left(A_j(t) - \hat{A}_j\right)^2 + 1}$$

(13)

where T is the number of iterations, $\hat{A}_j$ is the actual output of jth concept of system, $A_j(t)$ is simulated output at t iteration by computing.

In the procedure, the algorithm needs to call function M_NHL. Moreover, if the fitness and the number of iterations are not satisfied with the max, the next chromosome is created by select method, mutation method and recombination method. In our experiments, a simple one-point crossover, random mutation and roulette wheel selection are applied. The parameters are chosen and set as Table 1.

Table 1. The parameters in the integrated algorithm.

Parameters	Values	Meanings
probability of recombination	0.9	probability of single-point crossover
probability of mutation	0.5	probability of random mutation
population_size	50	the number of chromosomes
max_generation	500,000	a maximum number of generations
max_fitness	[0.6, 0.9]	fitness thresholds
a	1000	a parameter in Equation (13)

The execution procedure of the algorithm integrating RCGA and NHL consisted of the following steps:

Step 1: Initialize the parameters by the Table 1
Step 2: Randomly initialize population_size chromosomes, $g = 0, t = 0$
Step 3: Repeat for each dimension j:
    3.1   Calculate $A_j(t)$ by Equation (11) and $\mathbf{w}_j$ based on M_NHL
Step 4: Repeat for each chromosome:
    4.1   Calculate the *fitness* by $A_j(t)$ and Equation (13)
Step 5: Get max of the *fitness* and the W
Step 6: if max of *fitness* not more than *max_fitness* and g not more than *max_generation*
    6.1   Select chromosomes by roulette wheel selection
    6.2   Recombination the chromosomes by single-point crossover
    6.3   Random mutation to the chromosomes by the probability
    6.4   Set $t = i + 1$, Repeat for each chromosome:

6.4.1   Calculate $A_j(t)$ by Equation (1)

6.5   Set $g = g + 1$, go to Step 5

Step 7:   The W is the optimal chromosome.

The criterion in Equation (14) is defined as a normalized average error between corresponding concept values at each iteration between the two states. The error is used to define the accuracy of the algorithm simulating FCM.

$$error = \frac{1}{M \times T} \sum_{t=0}^{T} \sum_{j=1}^{M} |A_j(t) - \hat{A}_j| \tag{14}$$

4. Results and Discussion

Experiments have been carried out using Mutagenesis describing molecular structure, which is a multi-relational dataset. The multi-relational system consists of five tables (relationships). They are Atom, Bond, Atom_1, Molatm and Mole, where Bond and Molatm play associations among other three tables. The class label is in Mole table. The Mutagenesis data describes 188 molecules falling in two classes, mutagenic (active) and non-mutagenic (inactive); 125 of these molecules are mutagenic.

A molecule is associated with multi-atoms through Molatm. An atom is associated with several atoms through Bond. So Mutagenesis can be expressed as a multi-relationship of three dimensions: Atom, Atom_1, and Mole. Each dimension has many fine-grained concepts, such as indl, inda, lumo, logp of molecular dimension.

There are three backgrounds of Mutagenesis shown in Table 2. The three multi-relational FCM structures as shown in Figures 3–5 are different in three different backgrounds. The dotted lines show high-level FCMs.

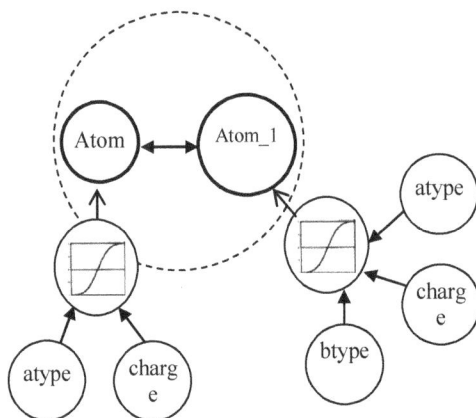

Figure 3. FCM structure of Mutagenesis in BK_0.

138

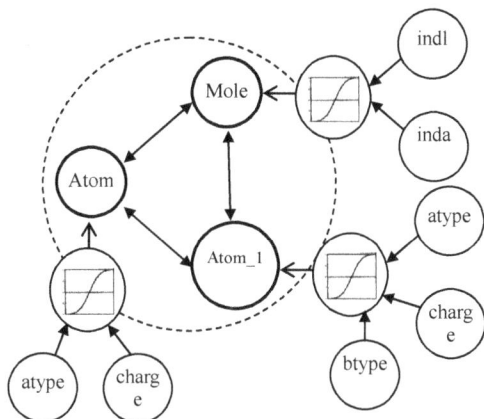

Figure 4. FCM structure of Mutagenesis in BK_1.

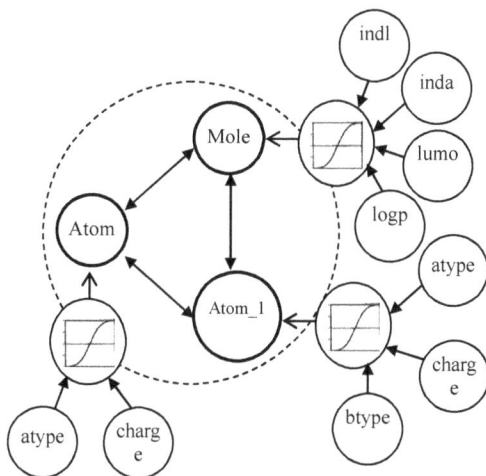

Figure 5. FCM structure of Mutagenesis in BK_2.

Table 2. Three kinds of background of Mutagenesis.

Background	Description
BK_0	Each compound includes the attributes of bond types, atom types, and partial charges on atoms
BK_1	Each compound includes indl and inda of mole besides those in BK_0
BK_2	Each compound includes all attributes that are logp and lumo of mole besides those in BK_1

The experiment is implemented for the class of molecular and the association weights among the three concepts (Atom, Atom_1 and Molecular) in the Mutagenesis.

For better operation efficiency in shorter runtime, an experiment, based on multi-relationship FCM and the integrated algorithm, has been carried out in the different fitness thresholds under three kinds of background. The learning runtimes are shown in the fitness thresholds from 0.6 to 0.9. From Figure 6, we can see that the changes of the runtime are not big under three kinds of background. The runtimes spent is changed. When the fitness threshold is at the interval of (0.65, 0.76], the operation takes less time.

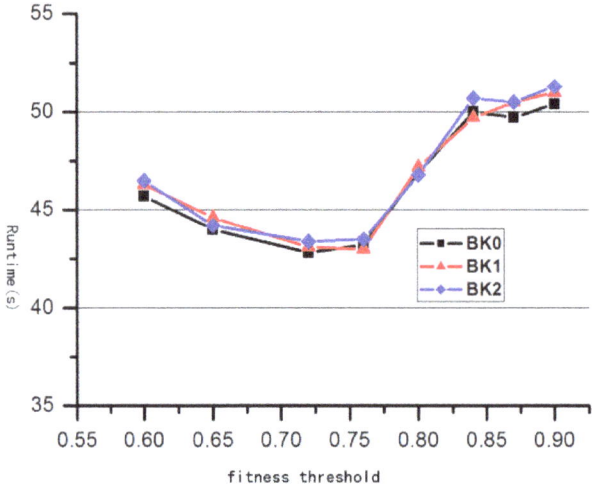

Figure 6. The learning runtime under the different fitness thresholds.

The classification results are compared in the *max_fitness* = 0.7 under three backgrounds. From the Table 3, the classification runtime of the integrated method is longer. The main reason for this is the costs in database access and FCM inference. The integrated method has better classification accuracy according to the label in Mole in three kinds of background knowledge; in particular, the accuracy rate is best in BK_1.

Table 3. Classification efficiency in different backgrounds.

Backgrounds	Runtime(s)	Accuracy (%)
BK_0	0.78	82.3%
BK_1	0.8	82.9%
BK_2	0.8	82.7%

And the method not only gets better classification, but also the association weights or causality for causal analysis of system, which is more than other methods.

For example, the association matrix of high-level FCM of Mutagenesis in BK_2 is shown in Figure 7.

$$
\begin{array}{c}
\begin{array}{ccc} \textbf{Atom} & \textbf{Atom\_1} & \textbf{Mole} \end{array} \\
\begin{array}{c} \textbf{Atom} \\ \textbf{Atom\_1} \\ \textbf{Mole} \end{array}
\begin{bmatrix}
0 & 0.294 & 0.6308 \\
0.526 & 0 & 0.1801 \\
0.2127 & 0.4758 & 0
\end{bmatrix}
\end{array}
$$

Figure 7. Association matrix of high-level FCM of Mutagenesis in BK_2.

5. Conclusions

We construct a kind of multi-levels and multi-dimensions FCM to automatic model complex systems directly from multi-relational data resources. The multi-relational FCM include two levels and some dimensions. In the FCM, one concept in high-level has a summary evaluation in a dimension, which is inferred by the transformation function of low-level FCM. It has been solved that the weight vector in low-level FCM is learned by extended NHL from multi-instances for the inference. For getting better classification and causality, RCGA has been used in learning the association weights in high-level FCM. Moreover, the integrating algorithm of NHL and RCGA has been applied in the compounds of molecular of Mutagenesis, which obtains better accuracy and knowledge for causal analysis.

Acknowledgments: This research was supported by the Natural Science Foundation of Hebei Province (No. F2014508028), the National Natural Science Foundation of China (No. 61202027), the Project of Construction of Innovative Teams and Teacher Career Development for Universities and Colleges Under Beijing Municipality (No. IDHT20150507), and the Scientific Research Base Development Program of the Beijing Municipal Commission of Education (TJSHG201310028014).

Author Contributions: Zhen Peng performed the multi-relational FCM model construction and the integrated algorithm. Zhen Peng, Lifeng Wu and Zhenguo Chen carried out experimental work. Zhen Peng prepared the manuscript.

Conflicts of Interest: The authors declare no conflict of interest.

References

1. Han, J.W.; Kamber, M. *Concept and Technology of Data Mining*; Machinery Industry Press: Beijing, China, 2007; pp. 373–383.
2. Spyropoulou, E.; de Bie, T.; Boley, M. Interesting pattern mining in multi-relational data. *Data Knowl. Eng.* **2014**, *28*, 808–849.
3. Yin, X.; Han, J.; Yang, J.; Yu, P.S. Efficient classification across multiple database relations: A CrossMine approach. *IEEE Trans. Knowl. Data Eng.* **2006**, *18*, 770–783.

4. Xu, G.; Yang, B.; Qin, Y.; Zhang, W. Multi-relational Naive Bayesian classifier based on mutual information. *J. Univ. Sci. Technol. Beijing* **2008**, *30*, 943–966.

5. Liu, H.; Yin, X.; Han, J. An efficient multi-relational naive Bayesian classifier based on semantic relationship graphs. In Proceedings of the 2005 ACM-SIGKDD Workshop on Multi-Relational Data Mining (KDD/MRDM05), Chicago, IL, USA, 21–24 August 2005.

6. Kosko, B. Fuzzy cognitive maps. *Int. J. Man Mach. Studs.* **1986**, *24*, 65–75.

7. Papageorgiou, E.I.; Salmeron, J.L. A review of fuzzy cognitive maps research during the last decade. *IEEE Trans. Fuzzy Syst.* **2013**, *12*, 66–79.

8. Groumpos, P.P. Large scale systems and fuzzy cognitive maps: A critical overview of challenges and research opportunities. *Annu. Rev. Control* **2014**, *38*, 93–102.

9. Zhang, Y. Modeling and Control of Dynamic Systems Based on Fuzzy Cognitive Maps. Ph.D. Thesis, Dalian University of Technology, Dalian, China, 2012.

10. Mago, V.K.; Bakker, L.; Papageorgiou, E.I.; Alimadad, A.; Borwein, P.; Dabbaghian, V. Fuzzy cognitive maps and cellular automata: An evolutionary approach for social systems modeling. *Appl. Soft Comput.* **2012**, *12*, 3771–3784.

11. Subramanian, J.; Karmegam, A.; Papageorgiou, E.; Papandrianos, N.I.; Vasukie, A. An integrated breast cancer risk assessment and management model based on fuzzy cognitive maps. *Comput. Methods Programs Biomed.* **2015**, *118*, 280–297.

12. Szwed, P.; Skrzynski, P. A new lightweight method for security risk assessment based on fuzzy cognitive maps. *Appl. Math. Comput. Sci.* **2014**, *24*, 213–225.

13. Peng, Z.; Peng, J.; Zhao, W.; Chen, Z. Research on FCM and NHL based High Order mining driven by big data. *Math. Probl. Eng.* **2015**, *2015*.

14. Papakostas, G.A.; Koulouriotis, D.E.; Polydoros, A.S.; Tourassis, V.D. Towards Hebbian learning of fuzzy cognitive maps in pattern classification problems. *Expert Syst. Appl.* **2012**, *39*, 10620–10629.

15. Salmeron, J.L.; Papageorgiou, E.I. Fuzzy grey cognitive maps and nonlinear Hebbian learning in process control. *Appl. Intell.* **2014**, *41*, 223–234.

16. Beena, P.; Ganguli, R. Structural damage detection using fuzzy cognitive maps and Hebbian learning. *Appl. Soft Comput.* **2011**, *11*, 1014–1020.

17. Kim, M.-C.; Kim, C.O.; Hong, S.R.; Kwon, I.-H. Forward-backward analysis of RFID-enabled supply chain using fuzzy cognitive map and genetic algorithm. *Expert Syst. Appl.* **2008**, *35*, 1166–1176.

18. Froelich, W.; Papageorgiou, E.I. Extended evolutionary learning of fuzzy cognitive maps for the prediction of multivariate time-series. *Fuzzy Cogn. Maps Appl. Sci. Eng.* **2014**, *2014*, 121–131.

19. Froelich, W.; Salmeron, J.L. Evolutionary learning of fuzzy grey cognitive maps for forecasting of multivariate, interval-valued time series. *Int. J. Approx. Reason.* **2014**, *55*, 1319–1335.

20. Wojciech, S.; Lukasz, K.; Witold, P. A divide and conquer method for learning large fuzzy cognitive maps. *Fuzzy Sets Syst.* **2010**, *161*, 2515–2532.

21. Sudjianto, A.; Hassoun, M.H. Statistical basis of nonlinear Hebbian learning and application to clustering. *Neural Netw.* **1995**, *8*, 707–715.

Dual-Tree Complex Wavelet Transform and Twin Support Vector Machine for Pathological Brain Detection

Shuihua Wang, Siyuan Lu, Zhengchao Dong, Jiquan Yang, Ming Yang
and Yudong Zhang

Abstract: (**Aim**) Classification of brain images as pathological or healthy case is a key pre-clinical step for potential patients. Manual classification is irreproducible and unreliable. In this study, we aim to develop an automatic classification system of brain images in magnetic resonance imaging (MRI). (**Method**) Three datasets were downloaded from the Internet. Those images are of T2-weighted along axial plane with size of 256×256. We utilized an s-level decomposition on the basis of dual-tree complex wavelet transform (DTCWT), in order to obtain $12s$ "variance and entropy (VE)" features from each subband. Afterwards, we used support vector machine (SVM) and its two variants: the generalized eigenvalue proximal SVM (GEPSVM) and the twin SVM (TSVM), as the classifiers. In all, we proposed three novel approaches: DTCWT + VE + SVM, DTCWT + VE + GEPSVM, and DTCWT + VE + TSVM. (**Results**) The results showed that our "DTCWT + VE + TSVM" obtained an average accuracy of 99.57%, which was not only better than the two other proposed methods, but also superior to 12 state-of-the-art approaches. In addition, parameter estimation showed the classification accuracy achieved the largest when the decomposition level s was assigned with a value of 1. Further, we used 100 slices from real subjects, and we found our proposed method was superior to human reports from neuroradiologists. (**Conclusions**) This proposed system is effective and feasible.

Reprinted from *Appl. Sci.* Cite as: Wang, S.; Lu, S.; Dong, Z.; Yang, J.; Yang, M.; Zhang, Y. Dual-Tree Complex Wavelet Transform and Twin Support Vector Machine for Pathological Brain Detection. *Appl. Sci.* **2016**, *6*, 169.

1. Introduction

Stroke, brain tumors, neurodegenerative diseases, and inflammatory/infectious diseases are the four main types of brain diseases. Stroke is also called vascular disease of cerebral circulation. Brain tumors occur when abnormal cells form inside the brain. Neurodegenerative diseases occur when neurons lose structure or function progressively. Inflammatory/infectious disease suffers from inflammation or infection in or around the brain tissues. All diseases cause serious problems for both patients and the society. Hence, it is important to make early diagnosis system, with the aim of providing more opportunities for better clinical trials. This type of task is commonly named as "pathological brain detection (PBD)".

Magnetic resonance imaging (MRI) offers the best diagnostic information in the brain; however, to make a diagnosis usually needs human manual interpretation. Existing manual methods are costly, tedious, lengthy, and irreproducible, because of the huge volume data of MRI. Those shortcomings lead to the necessity to develop automatic tools such as computer-aided diagnosis (CAD) systems [1,2]. Due to the better performance provided by magnetic resonance (MR) images, many CAD systems are based on MR images [3].

Existing methods over brain CAD systems could be divided into two types according to the data dimension. One type is for three-dimensional (3D) image, but it needs to scan the whole brain. The other type is based on a single-slice that contains the disease related areas, which is cheap and commonly used in Chinese hospitals. El-Dahshan *et al.* [4] employed a 3-level discrete wavelet transform (DWT), followed by principal component analysis (PCA) to reduce features. Finally, they used K-nearest neighbor (KNN) for classification. Patnaik *et al.* [5] utilized DWT to extract the approximation coefficients. Then, they employed support vector machine (SVM) for classification. Dong *et al.* [6] further suggested to train feedforward neural network (FNN) with a novel scaled conjugate gradient (SCG) approach. Their proposed classification method achieved good results in MRI classification. Wu [7] proposed to use kernel SVM (KSVM), and suggested three new kernels: Gaussian radial basis, homogeneous polynomial, and inhomogeneous polynomial. Das *et al.* [8] proposed to combine Ripplet transform (RT) and PCA and least square SVM (LS-SVM), and the 5×5 cross validation test showed high classification accuracies. El-Dahshan *et al.* [9] used the feedback pulse-coupled neural network to preprocess the MR images, the DWT and PCA for features extraction and reduction, and the FBPNN to detect pathological brains from normal brains. Dong *et al.* [10] combined discrete wavelet packet transform (DWPT) and Tsallis entropy (TE). In order to segment and classify malignant and benign brain tumor slices in Alzheimer's disease (AD), Wang *et al.* [11] employed stationary wavelet transform (abbreviated as SWT) to replace the common used DWT. Besides, they proposed a hybridization of Particle swarm optimization and Artificial bee colony (HPA) algorithm to obtain the optimal weights and biases of FNN. Nazir *et al.* [12] implemented denoising at first. Their method achieved an overall accuracy of 91.8%. Sun *et al.* [13] combined wavelet entropy (WE) and Hu moment invariant (HMI) as features.

After analyzing the above methods, we found all literatures treated the PBD problem as a classification problem and their studies aimed at improving the classification accuracy. As we know, a standard classification task is composed of two steps: feature extraction and classifier training.

The former step "feature extraction" in PBD usually employed discrete wavelet transform (DWT) [5]. The reason is DWT can provide multiresolution analysis at any desired scale for a particular brain image. Besides, the abundant texture features in various brain regions are coherent with wavelet analysis [14,15]. However, DWT is shift variant, *i.e.*, a slight shift in the image degrades the performance of DWT-based classification. Our study chose a variant of DWT, viz., the dual-tree complex wavelet transform (DTCWT). Scholars have proven that the DTCWT offers "*more directional selectivity*" than canonical DWT, with merely 2^n redundancy for data of n-dimensional [16–18].

Although stationary wavelet transform (SWT) can also deal with the shift variance problem, it will lead to more redundancy than DTCWT. Then, we need to extract features from the DTCWT results. In this paper, we proposed to use variance and entropy (VE) [19] from all the subbands of DTCWT. Although energy is also a common feature extracted from the wavelet subbands, scholars have proven it is not as efficient as entropy in MR image classification [20,21]. Besides, variance with the form of $E[(x - \mu)^2]$ (E, expectation; x, random variable; μ, the mean) indicates how data points be close to the mean value, while energy with the form of $E[x^2]$ does not consider the mean value. Hence, it is self-explanatory that variance will perform better than energy even if the expected mean value has a slight shift.

The latter step "classifier training" usually employed support vector machines (SVMs) or artificial neural network. Compared with other conventional classification methods, SVMs have significant advantages of elegant mathematical tractability [22], direct geometric interpretation [23], high accuracy [24], *etc.* Hence, we continued to choose SVM. Besides, two variants of SVM were introduced in this study: generalized eigenvalue proximal SVM (GEPSVM) and twin SVM (TSVM), with the aim of augmenting the classification performance further.

The contribution of our study is three-fold: We applied DTCWT to pathological brain detection. We applied both variance and entropy to extract features. We applied TSVM for classification. The structure of the paper is organized as follows: Section 2 contains the materials used in this study. Section 3 presents the dual-tree complex wavelet transform, and offers the mathematical fundamental of SVM and its two variants. Section 4 designs the experiment and gives the evaluation measures. Section 5 contains the experimental results and offers the discussions. Section 6 is devoted to conclusions. The abbreviations used in this work are listed in the end of this paper.

2. Materials

At present, there are three benchmark datasets as Dataset66, Dataset160, and Dataset255, of different sizes of 66-, 160-, and 255-images, respectively. All datasets contain T2-weighted MR brain images obtained along axial plane with size of

256 × 256. We downloaded all the slices of subjects from the website of Medical School of Harvard University (Boston, MA, USA) [25]. Then, we selected five slices from each subject. The selection criterion is that for healthy subjects, these slices were selected at random. For pathological subjects, the slices should contain the lesions by confirmation of thee radiologists with ten years of experiences.

The former two datasets (Dataset66 & Dataset160) consisted of 7 types of diseases (meningioma, AD, AD plus visual agnosia, sarcoma, Pick's disease, Huntington's disease, and glioma) along with normal brain images. The last dataset "Dataset255" contains all 7 types of diseases as mentioned before, and 4 new diseases (multiple sclerosis, chronic subdural hematoma, herpes encephalitis, and cerebral toxoplasmosis).

Figure 1 shows samples of brain MR images. Our method is for hospital other than research. In Chinese hospitals, we usually scanned one slice that is closest to the potential focus, other than the whole brain. Hence, one slice was obtained from one subject. Each slice in Figure 1 is selected from regions related to the foci of diseases (in total 26 axial slices).

Note that we treated all different diseased brains as pathological, so our task is a two-class classification problem, that is, to detect pathological brains from healthy brains. The whole image is treated as the input. We did not choose local characteristics like point and edge, and we extract global image characteristics that are further learned by the CAD system. Note that our method is different from the way neuroradiologists do. They usually select local features and compare with standard template to check whether focuses exist, such as shrink, expansion, bleeding, inflammation, *etc.* While our method is like AlphaGo [26], the computer scientists give the machine enough data, and then the machine can learn how to classify automatically.

Including subjects' information (age, gender, handedness, memory test, *etc.*) can add more information, and thus may help us to improve the classification performance. Nevertheless, this CAD system in our study is only based on the imaging data. Besides, the imaging data from the website does not contain the subjects' information.

The cost of predicting pathological to healthy is severe; because the patients may be told that, she/he is healthy and thus ignore the mild symptoms displayed. The treatments of patients may be deferred. Nevertheless, the cost of misclassification of healthy to pathological is low, since correct remedy can be given by other diagnosis means.

This cost-sensitivity (CS) problem was solved by changing the class distribution at the beginning state, since original data was accessible. That means, we intentionally picked up more pathological brains than healthy ones into the dataset, with the aim of making the classifier biased to pathological brains, to solve the CS problem [27]. The overfitting problem would be monitored by cross validation technique.

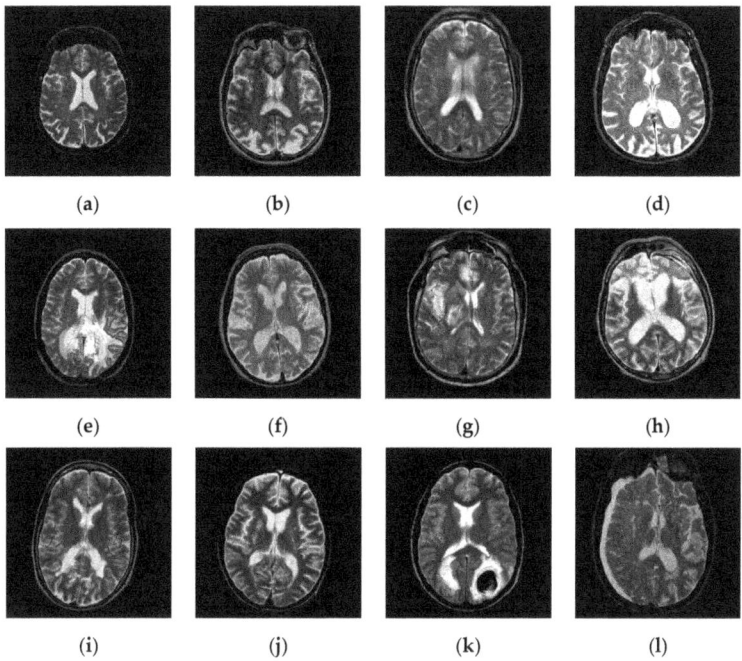

Figure 1. Sample of pathological brains. (**a**) Normal brain; (**b**) Alzheimer's disease (AD) with visual agnosia; (**c**) Meningioma; (**d**) AD; (**e**) Glioma; (**f**) Huntington's disease; (**g**) Herpes encephalitis; (**h**) Pick's disease; (**i**) Multiple sclerosis; (**j**) Cerebral toxoplasmosis; (**k**) Sarcoma; (**l**) Subdural hematoma.

3. Methodology

The proposed method consists of three decisive steps: wavelet analysis by dual-tree complex wavelet transform (DTCWT), feature extraction by "Variance & Entropy (VE)", and classification by three independent classifiers (SVM, GEPSVM, and TSVM). Figure 2 illustrates our modular framework. The output of DTCWT is wavelet subband coefficients, which are then submitted to VE block.

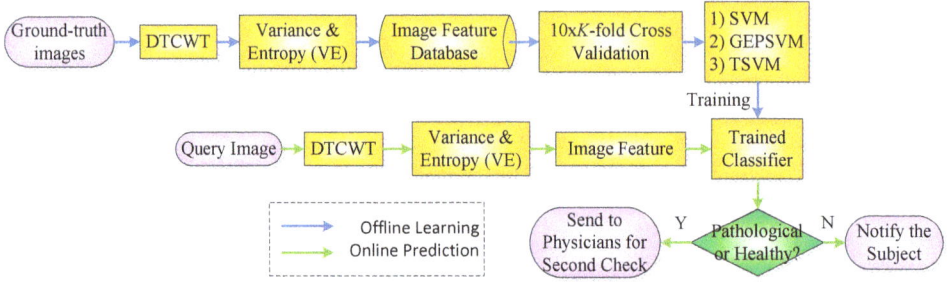

Figure 2. Modular framework of the proposed system for magnetic resonance (MR) brain classification (*K* may be 5 or 6 according to the dataset).

3.1. Discrete Wavelet Transform

The discrete wavelet transform (DWT) is an image processing method [28] that provides multi-scale representation of a given signal or image [29]. Standard DWT is vulnerable to shift variance problem, and only has horizontal and vertical directional selectivity [30]. Suppose *s* represents a particular signal, *n* represents the sampling point, *h* and *g* represents a high-pass filter and low-pass filter, respectively, *H* and *L* represents the coefficients of high-pass and low-pass subbands. We have

$$H(n) = \sum_m h(2n - m)s(m) \tag{1}$$

$$L(n) = \sum_m g(2n - m)s(m) \tag{2}$$

Figure 3 shows the directional selectivity of DWT. The *LH* denotes a low-pass filter along *x*-axis and high-pass filter along *y*-axis. *HL* denotes a high-pass filter along *x*-axis followed by a low-pass filter along *y*-axis. The *LL* denotes low-pass filters along both directions, and *HH* denotes high-pass filters along both directions.

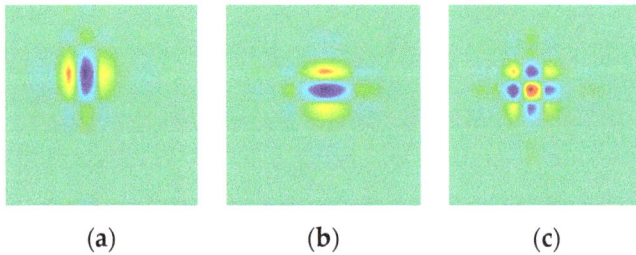

(a)   (b)   (c)

Figure 3. Directional Selectivity of discrete wavelet transform (DWT). (*L* = Low, *H* = High). (**a**) *LH*; (**b**) *HL*; (**c**) *HH*.

Here the *HL* and *LH* have well-defined for both vertical and horizontal orientations. For the *HH*, it mixes directions of both −45 and +45 degrees together, which stems from the use of real-valued filters in DWT. This mixing also impedes the direction check [31].

3.2. Dual-Tree Complex Wavelet Transform

To help improve the directional selectivity impaired by DWT, we proposed a dual-tree DWT, which was implemented by two separate two-channel filter banks. Note that the scaling and wavelet filters in the dual-tree cannot be selected arbitrarily [32]. In one tree, the wavelet and scaling filters should produce a wavelet and scaling function, which are approximate Hilbert transforms of those generated by another tree [33]. In this way, the wavelet generated from both trees and the complexed-valued scaling function are approximately analytic, and are called dual-tree complex wavelet transform (DTCWT).

DTCWT obtained directional selectivity by using approximately analytic wavelets, *i.e.*, they have support on only one half of the whole frequency domain [34]. At each scale of a 2D DTCWT, it produces in total six directionally selective subbands ($\pm 15°$, $\pm 45°$, $\pm 75°$) for both real ($\mathbb{R}$) and imaginary ($\mathbb{I}$) parts [35]. Figure 4 shows the directional selectivity of DTCWT. The first row depicts the 6 directional wavelets of the real oriented DTCWT, and the second row shows the imaginary counterpart. The $\mathbb{R}$ and $\mathbb{I}$ parts are oriented in the same direction, and they together form the DTCWT as

$$\mathbb{M} = \sqrt{\mathbb{R}^2 + \mathbb{I}^2} \tag{3}$$

where $\mathbb{M}$ represents the magnitude of the DTCWT coefficients.

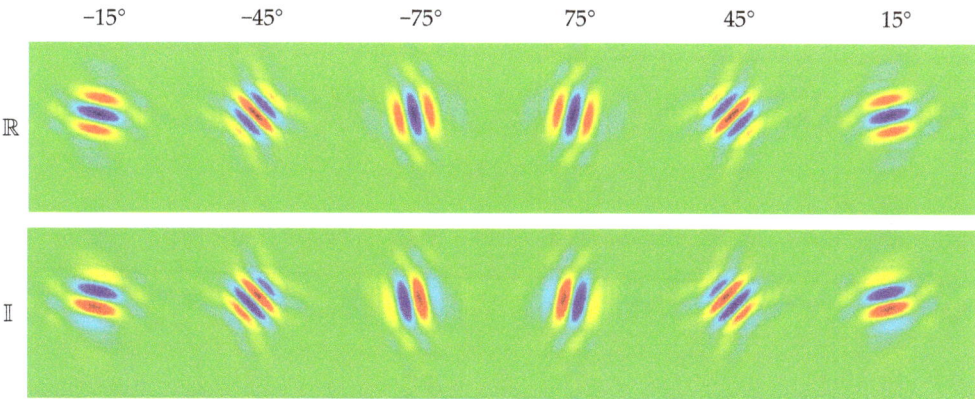

Figure 4. Directional Selectivity of dual-tree complex wavelet transform (DTCWT). ($\mathbb{R}$ = Real, $\mathbb{I}$ = Imaginary).

149

3.3. Comparison between DWT and DTCWT

To compare the directional selectivity ability between DWT and DTCWT, we performed a simulation experiment. Two simulation images (a pentagon and a heptagon) were generated. We decomposed both images to 4-level by DWT and DTCWT, respectively. Then, we reconstructed them to obtain an approximation to original images by a 4-th level detail subband. The results were shown in Figure 5.

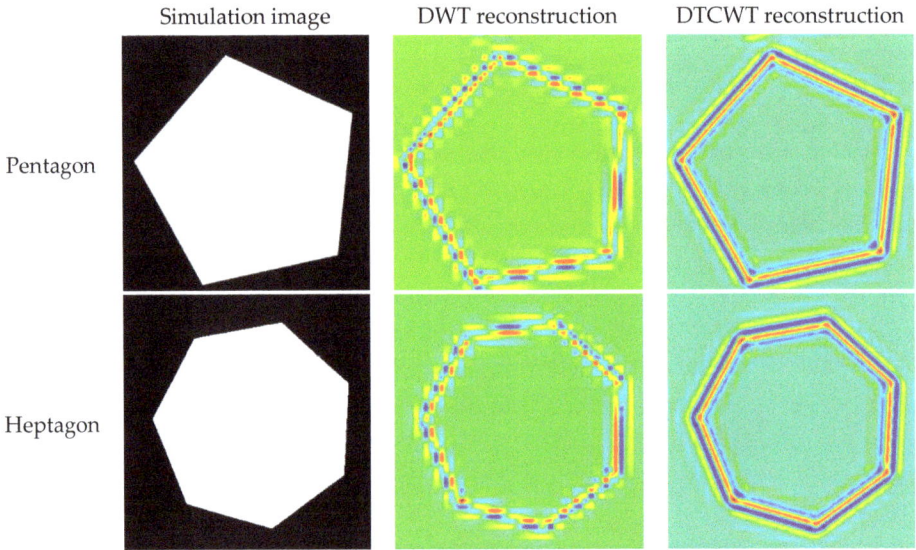

Figure 5. The reconstruction comparison between DWT and DTCWT.

The first column in Figure 5 shows the simulation images, the second column shows the DWT reconstruction results, and the last column shows the DTCWT reconstruction results. Both DWT and DTCWT can extract edges from detail subbands, which are abundant in brain tissues.

Those edges are discriminant features that are different in pathological brains and healthy brains. The reason is all the focus-related areas contain either shrink, or expand, or bleed, or become inflamed. Those that will yield structural alternations that are associated with edges. We find from the last column in Figure 5 that the edges detected by DTCWT have a clear contour, so DTCWT can detect nearly all directions clearly and perfectly. Nevertheless, the edges detected by DWT (See Figure 5) are discontinued, stemming from that DWT can only detect horizontal and vertical edges. The results fall in line with Figure 3.

3.4. Variance and Entropy (VE)

Based on the coefficients of DTCWT, we extract variance and entropy (VE) features for each decomposition level s and each direction d. Suppose (x, y) is the spatial coordinate of the corresponding subband, and (L, W) is the length and width of the corresponding subband, we can define the variance $V_{(s, d)}$ as

$$V_{(s,d)} = \frac{1}{LW} \sum_{x=1}^{L} \sum_{y=1}^{W} \left(\mathbb{M}_{(s,d)}(x,y) - \mu_{(s,d)} \right)$$

$$\mu_{(s,d)} = \frac{1}{LW} \sum_{x=1}^{L} \sum_{y=1}^{W} \mathbb{M}_{(s,d)}(x,y)$$

(4)

here μ denotes the mathematical expectation of $\mathbb{M}$. The variance V measures the spread of grey-level distribution of the subbands. The larger the value of V is, the more widely the gray-levels of the image vary. V also reflect the contrast of the texture.

Another indicator is the entropy E, which measures the randomness of the gray-level distribution [36]. The larger the value of V is, the more randomly the gray-level distribution spreads [37]. The entropy E is defined in following form:

$$E_{(s,d)} = -\frac{1}{LW} \sum_{x=1}^{L} \sum_{y=1}^{W} \mathbb{P}\left(\mathbb{M}_{(s,d)}(x,y) \right) \log \mathbb{P}\left(\mathbb{M}_{(s,d)}(x,y) \right)$$

(5)

Here, $\mathbb{P}$ denotes the probability function. Both variance and entropy are sufficient to produce a good performance. All directional subbands of those two kinds of features are combined to form a new feature set $\mathbf{V}_s$ and $\mathbf{E}_s$ as

$$\mathbf{V}_s = \frac{[V_{(s,-15)}, V_{(s,15)}, V_{(s,-45)}, V_{(s,45)}, V_{(s,-75)}, V_{(s,75)}]}{\sqrt{V_{(s,-15)}^2 + V_{(s,15)}^2 + V_{(s,-45)}^2 + V_{(s,45)}^2 + V_{(s,-75)}^2 + V_{(s,75)}^2}}$$

(6)

$$\mathbf{E}_s = \frac{[E_{(s,-15)}, E_{(s,15)}, E_{(s,-45)}, E_{(s,45)}, E_{(s,-75)}, E_{(s,75)}]}{\sqrt{E_{(s,-15)}^2 + E_{(s,15)}^2 + E_{(s,-45)}^2 + E_{(s,45)}^2 + E_{(s,-75)}^2 + E_{(s,75)}^2}}$$

(7)

Hence, we extract 12 features for each scale, among which 6 is for $\mathbf{V}_s$ and 6 for $\mathbf{E}_s$. For an s-level decomposition, we totally obtain $12s$ features VE as

$$\mathbf{VE} = [\mathbf{V_1, V_2, ..., V_s, E_1, E_2, ..., E_s}]$$

(8)

The $12s$ **VE**s were then submitted into classifiers. SVM is now probably treated as one of the most excellent classification approach in small-size (less than one-thousand samples) problem [7]. To further enhance the classification performance, two new variants of SVM were introduced:

3.5. Generalized Eigenvalue Proximal SVM

Mangasarian and Wild [38] proposed the generalized eigenvalue proximal SVM (GEPSVM). It drops the parallelism condition on the two hyperplanes (remember the parallelism is necessary in original SVM). Latest literatures showed that GEPSVM yielded superior classification performance to canonical support vector machines [39,40].

Suppose samples are from either class 1 (denote by symbol X_1) or class 2 (denoted by symbol X_2), respectively. The GEPSVM finds the two optimal nonparallel planes with the form of ($\mathbf{w}$ and b denotes the weight and bias of the classifier, respectively)

$$\mathbf{w}_1^T x - b_1 = 0 \text{ and } \mathbf{w}_2^T x - b_2 = 0 \tag{9}$$

To obtain the first plane, we deduce from Equation (9) and get the following solution

$$(\mathbf{w}_1, b_1) = \underset{(\mathbf{w},b) \neq 0}{\arg \min} \frac{||\mathbf{w}^T X_1 - o^T b||^2 / ||z||^2}{||\mathbf{w}^T X_2 - o^T b||^2 / ||z||^2} \tag{10}$$

$$z \leftarrow \begin{bmatrix} \mathbf{w} \\ b \end{bmatrix} \tag{11}$$

where o is a vector of ones of appropriate dimensions. Simplifying formula (10) gives

$$\underset{(\mathbf{w},b) \neq 0}{\min} \frac{||\mathbf{w}^T X_1 - o^T b||^2}{||\mathbf{w}^T X_2 - o^T b||^2} \tag{12}$$

We include the Tikhonov regularization to decrease the norm of z, which corresponds to the first hyperplane. The new equation including Tikhonov regularization term is:

$$\underset{(\mathbf{w},b) \neq 0}{\min} \frac{||\mathbf{w}^T X_1 - o^T b||^2 + t||z||^2}{||\mathbf{w}^T X_2 - o^T b||^2} \tag{13}$$

where t is a positive (or zero) Tikhonov factor. Formula (13) turns to the "Rayleigh Quotient (RQ)" in the following form of

$$z_1 = \underset{z \neq 0}{\arg \min} \frac{z^T P z}{z^T Q z} \tag{14}$$

where P and Q are symmetric matrices in $\mathbb{R}^{(p+1)\times(p+1)}$ in the forms of

$$P \leftarrow \begin{bmatrix} X_1 & -o \end{bmatrix}^T \begin{bmatrix} X_1 & -o \end{bmatrix} + tI \tag{15}$$

$$Q \leftarrow \begin{bmatrix} X_2 & -o \end{bmatrix}^T \begin{bmatrix} X_2 & -o \end{bmatrix} \tag{16}$$

Solution of (14) is deduced by solving a generalized eigenvalue problem, after using the stationarity and boundedness characteristics of RQ.

$$Pz = \lambda Qz, z \neq 0 \tag{17}$$

Here the optimal minimum of (14) is obtained at an eigenvector z_1 corresponding to the smallest eigenvalue λ_{min} of formula (17). Therefore, $\mathbf{w}_1$ and b_1 can be obtained through formula (11), and used to determine the plane in formula (9). Afterwards, a similar optimization problem is generated that is analogous to (12) by exchanging the symbols of X_1 and X_2. $z_2{}^*$ can be obtained in a similar way.

3.6. Twin Support Vector Machine

In 2007, Jayadeva et al. [41] presented a novel classifier as twin support vector machine (TSVM). The TSVM is similar to GEPSVM in the way that both obtain non-parallel hyperplanes. The difference lies in that GEPSVM and TSVM are formulated entirely different. Both quadratic programming (QP) problems in TSVM pair are formulated as a typical SVM. Reports have shown that TSVM is better than both SVM and GEPSVM [42–44]. Mathematically, the TSVM is constructed by solving the two QP problems

$$\min_{w_1,b_1,q} \tfrac{1}{2} (X_1\mathbf{w}_1 + o_1b_1)^T (X_1\mathbf{w}_1 + o_1b_1) + c_1 o_2^T q \tag{18}$$
$$\text{s.t.} \ -(X_2\mathbf{w}_1 + o_2b_1) + q \geqslant o_2, q \geqslant 0$$

$$\min_{w_2,b_2,q} \tfrac{1}{2} (X_2\mathbf{w}_2 + o_2b_2)^T (X_2\mathbf{w}_2 + o_2b_2) + c_2 o_1^T q \tag{19}$$
$$\text{s.t.} \ -(X_1\mathbf{w}_2 + o_1b_2) + q \geqslant o_1, q \geqslant 0$$

here q is a nonnegative slack variance. c_i ($i = 1,2$) are positive parameters, and o_i ($i = 1,2$) is the same as in formula (10). By this mean, the TSVM constructed two hyperplanes [45]. The first term in equations of (18) and (19) is the sum of squared distances. The second one represents the sum of error variables. Therefore, minimizing Equations (18) and (19) will force the hyperplanes approximate to data in each class, and minimize the misclassification rate [46]. Finally, the constraint requires the hyperplane to be at a distance of more than one from points of the other class. Another advantage of TSVM is that its convergence rate is four times faster than conventional SVM [47].

3.7. Pseudocode of the Whole System

The implementation covers two phases: offline learning and online prediction, with the goal of training the classifier and prediction of new instances, respectively. Table 1 offers the pseudocode of proposed methods.

Table 1. Pseudocode of our system.

		Phase I: Offline learning
Step A	Wavelet Analysis	Perform s-level dual-tree complex wavelet transform (DTCWT) on every image in the ground-truth dataset
Step B	Feature Extraction	Obtain $12 \times s$ features ($6 \times s$ Variances and $6s$ Entropies, and s represents the decomposition level) from the subbands of DTCWT
Step C	Training	Submit the set of features together with the class labels to the classifier, in order to train its weights/biases.
Step D	Evaluation	Record the classification performance based on a $10 \times K$ - fold stratified cross validation.
		Phase II: Online prediction
Step A	Wavelet Analysis	Perform s-level DTCWT on the real query image (independent from training images)
Step B	Feature Extraction	Obtain VE feature set
Step C	Prediction	Feed the VE feature set into the trained classifier, and obtain the output.

4. Experiment Design

4.1. Statistical Setting

In order to carry out a strict statistical analysis, stratified cross validation (SCV) was used since it is a model validation technique for small-size data [48]. 6-fold SCV was employed for Dataset66, and 5-fold SCV was employed for Dataset160 and Dataset255. The SCV setting was listed in Table 2.

Table 2. Stratified cross validation (SCV) setting of all datasets.

Dataset	No. of Fold	Training		Validation		Total	
		H	P	H	P	H	P
Dataset66	6	15	40	3	8	18	48
Dataset160	5	16	112	4	28	20	140
Dataset255	5	28	176	7	44	35	220

(H = Healthy, P = Pathological).

The 10-fold is not used because of two reasons: One is that past literatures used the same setting as Table 2. Another is for stratification, viz., we expect to guarantee each fold covers the same class numbers. If we divide the dataset into 10-folds, then the stratification will be breached.

Figure 6 illustrates an example of K-fold SCV, by which the dataset is partitioned into K folds with the same class distributions. The (K-1) folds are used for training, and the rest fold for test, *i.e.*, query images come from the rest fold. The evaluation is based on test images. This above process repeats K times so that each fold is used as test once. The final accuracy of K-fold SCV is obtained by averaging the K results. The K-fold SCV repeats 10 times to further remove randomness (See Section 5.4).

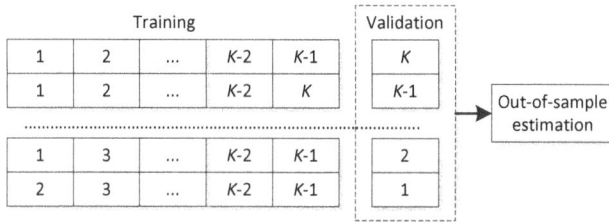

Training					Validation
1	2	...	K-2	K-1	K
1	2	...	K-2	K	K-1
1	3	...	K-2	K-1	2
2	3	...	K-2	K-1	1

Out-of-sample estimation

Figure 6. A K-fold SCV.

4.2. Parameter Estimation for s

It remains an issue of finding the optimal value of decomposition level s. From the view of information provided, a smaller s offers less information and a larger s offers more information to the classifier. From the view of avoiding overfit, a smaller s may prevent overfitting in greater degree than a larger s. This study used grid-search [49] method to find the optimal value of s, *i.e.*, we vary the value of s from 1 to 5 with increment of 1, and check the corresponding average accuracies. The one associated with the largest accuracy is the optimal value of s.

4.3. Evaluation

The pathological brains are treated as positive, while the healthy brains as negative. To evaluate the performance, we first calculated overall confusion matrix of 10 runs, then calculate the TP (True Positive), TN (True Negative), FP (False Positive), and FN (False Negative). The pathological brains were set to true and healthy ones to false, following common convention. The classification accuracy (Acc) is defined as:

$$Acc = \frac{TP + TN}{TP + TN + FP + FN} \tag{20}$$

5. Results and Discussions

The algorithms were in-house developed based on 64 bit Matlab 2015a (The Mathworks ©, Natick, MA, USA). Figure 7 shows the graphical user interface (GUI). The simulation experiments were implemented on the platform of P4 IBM with 3.2 GHz processor, and 8 GB random access memory (RAM), running under Windows 7 operating system.

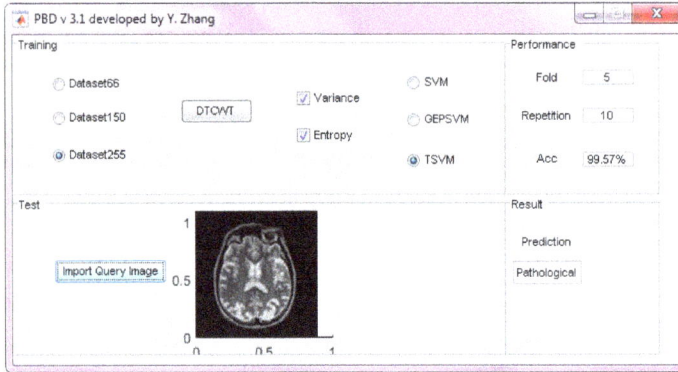

Figure 7. Graphical user interface (GUI) of our developed programs.

5.1. Classifier Comparison

In the second experiment, we compared three classifiers: SVM, GEPSVM, and TSVM. All three datasets are tested. A 10 runs of k-fold SCV was carried out. Accuracy was used for evaluation. The results are listed in Table 3. The SVM achieved 100.00%, 99.69%, and 98.43% accuracy for Dataset66, Dataset160, and Dataset255, respectively. The GEPSVM achieved accuracy of 100.00%, 99.75%, and 99.25% for three datasets. The TSVM yielded accuracy of 100.00%, 100.00%, and 99.57%, respectively.

Table 3. Accuracy Comparison based on 10 runs of k-fold SCV (Unit: %).

Our Methods	Dataset66	Dataset160	Dataset255
DTCWT + VE + SVM	100.00	99.69	98.43
DTCWT + VE + GEPSVM	100.00	99.75	99.25
DTCWT + VE + TSVM	100.00	100.00	99.57

Data in Table 3 indicate that GEPSVM is superior to standard SVM. For Dataset160, the *acc* of GEPSVM is higher than that of SVM by 0.06%. For Dataset255, the *acc* of GEPSVM is higher than that of SVM by 0.82%. Meanwhile, TSVM is superior to GEPSVM. For Dataset160, the *acc* of TSVM is 0.25 higher than that of GEPSVM. For Dataset255, the *acc* of TSVM is 0.32% higher than that of GEPSVM.

The parallel hyperplane setting restrains standard SVM to generate complicated and flexible hyperplanes. GEPSVM and TSVM discard this setting, so their performances are much better than SVM. TSVM is similar to GEPSVM in spirit, since both use non-parallel hyperplanes. The difference between them is TSVM uses simpler formulation than GEPSVM, and the former can be solved by merely two QP problems. Our results align with the finding in Kumar and Gopal [50], which says "*generalization performance of TSVM is better than GEPSVM and conventional SVM*". In following experiments, TSVM is the default classifier

5.2. Optimal Decomposition Level Selection

The value of decomposition level s was set in the range of (1, 2, 3, 4, 5). We chose the TSVM. All datasets were tested. A 10 runs of k-fold SCV was implemented with varying s. The curve of average accuracy *versus* against decomposition level is shown in Figure 8.

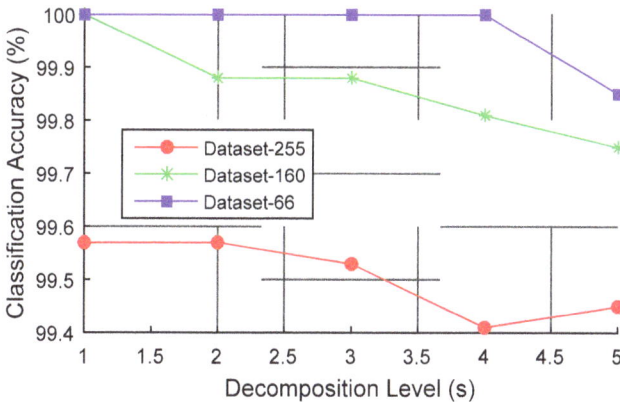

Figure 8. Classification Accuracy *versus* Decomposition Level (s).

Remember for $s = 1$, only 12 features are used. For $s = 2$, in total 24 features are used. The number of employed features is 12 times of the value of s. Figure 8 shows the relationship between *Acc versus s*. Here we find the *acc* has a tendency of decrease when the decomposition level s increases. The reason is more features will attenuate the classification performance [51]. Reducing the number of features can simplify the model, cost shorter training time, and augment generalization performance through reduction of variance [52].

5.3. Comparison to State-of-the-Art Approaches

We have already compared the SVM with its variants in Section 5.1. In this section, we compared the best of the proposed methods (DTCWT + VE + TSVM) with 12 state-of-the-art methods: DWT + PCA + KNN [4], DWT + SVM + RBF [5], DWT + PCA + SCG-FNN [6], DWT + PCA + SVM + RBF [7], RT + PCA + LS-SVM [8], PCNN + DWT + PCA + BPNN [9], DWPT + TE + GEPSVM [10], SWT + PCA + HPA-FNN [11], WE + HMI + GEPSVM [13], SWT + PCA + GEPSVM [53], FRFE + WTT + SVM [54], and SWT + PCA + SVM + RBF [55]. The meaning of these abbreviations can be found in the Abbreviations Section. The accuracy results were extracted directly from above literatures. The comparison is based on results obtained in each individual study.

Table 4 shows the comparison results between the best proposed method "DTCWT + VE + TSVM" with the state-of-the-art approaches. The first column lists the method name, the second column the number of features employed, the third column the total run times (all algorithms run 10 times, except some old algorithms ran five times which were reported in literature [8]), and the last three columns the average *acc* over three datasets.

Table 4. Classification comparison.

Algorithms	Feature #	Run #	*Acc*		
			Dataset66	Dataset160	Dataset255
DWT + PCA + KNN [4]	7	5	98.00	97.54	96.79
DWT + SVM + RBF [5]	4761	5	98.00	97.33	96.18
DWT + PCA + SCG-FNN [6]	19	5	100.00	99.27	98.82
DWT + PCA + SVM + RBF [7]	19	5	100.00	99.38	98.82
RT + PCA + LS-SVM [8]	9	5	100.00	100.00	99.39
PCNN + DWT + PCA + BPNN [9]	7	10	100.00	98.88	98.24
DWPT + TE + GEPSVM [10]	16	10	100.00	100.00	99.33
SWT + PCA + HPA-FNN [11]	7	10	100.00	100.00	99.45
WE + HMI + GEPSVM [13]	14	10	100.00	99.56	98.63
SWT + PCA + GEPSVM [53]	7	10	100.00	99.62	99.02
FRFE + WTT + SVM [54]	12	10	100.00	99.69	98.98
SWT + PCA + SVM + RBF [55]	7	10	100.00	99.69	99.06
DTCWT + VE + TSVM (Proposed)	12	10	100.00	100.00	99.57

(# represents number)

After investigating the results in Table 4, it is clear that 11 out of 13 methods achieve perfect classification (100%) over Dataset66, which stems from its small size. For a larger dataset (Dataset160), only four methods yield perfect classification. They are RT + PCA + LS-SVM [8], DWPT + TE + GEPSVM [10], SWT + PCA + HPA-FNN [11], and the proposed "DTCWT + VE + TSVM". A common point among the four methods is that they all used advanced feature extraction (RT, DWPT, SWT, and DTCWT) and classification techniques (LS-SVM, GEPSVM, HPA-FNN, and

TSVM). This suggests us to learn and apply latest advanced artificial-intelligence and machine-learning approaches to the field of MR brain classification. For the largest dataset (Dataset255), no algorithm achieves perfect classification, because there are relatively various types of diseases. Among all methods, this proposed "DTCWT + VE + TSVM" achieves the highest accuracy of nearly 99.57%, which demonstrates its effectiveness and feasibility.

It is reasonable that all methods achieved high accuracy. Retrospect a similar problem of facial recognition system (FRS), the latest FRS achieved nearly perfect performance and been applied for banking customers [56], vehicle security [57], *etc.* The pathological detection is simpler than face detection, because it does not need to identify each subject but identify the status (pathological or healthy). Hence, it is expected that our methods can achieve high classification accuracy.

The difference of accuracy in Table 4 is not significant, however, it is obtained by strict statistical analysis, viz., 10 runs of K-fold stratified cross validation. Hence, this slight improvement is reliable and convincing. Even the largest dataset only contains 255 images, so we will try to create a larger dataset that contains more images and more types of diseases.

The proposed CAD system cannot give physical meanings of particular brain regions. Nevertheless, after comparing classifier and human brains, we believe expert systems are similar to declarative memory, while support vector machines are similar to nondeclarative memory. Thus, it is impossible for SVMs (its variants) to give physical meanings. In the future, we may try to use expert systems that can mimic the reasoning process of doctors, but may not give as high accuracies as SVMs.

5.4. Results of Different Runs

The correct classification instance numbers, together with their accuracies, are listed in Table 5. In the table, each row lists the results of different runs, and each column lists the results of different folds. The last row averages the results, and the last column summarizes the results over different folds.

5.5. Computation Time

The computation time of each step of our method was calculated and recorded. The training time was recorded over Dataset255. The results of offline-learning and online-prediction are listed in Tables 6 and 7, respectively.

The computation time results in Table 6 provides that DTCWT costs 8.41 s, VE costs 1.81 s, and TSVM training costs 0.29 s, in the offline-learning procedure. This is because there are 255 images in the dataset, and the training process need to handle all images. The total time is 10.51 s.

The online-prediction time only deals with one query image, so the computation time reduces sharply. Table 7 provides that DTCWT costs 0.037 s, VE costs 0.009 s, and TSVM costs 0.003 s for one query image. Its total time is 0.049 s. Therefore, the proposed system is feasible in practice.

Table 5. Detailed Results of DTCWT + VE + TSVM over Dataset255.

Run	F1	F2	F3	F4	F5	Total
1	51(100.00%)	51(100.00%)	51(100.00%)	51(100.00%)	51(100.00%)	255(100.00%)
2	51(100.00%)	50(98.04%)	51(100.00%)	51(100.00%)	51(100.00%)	254(99.61%)
3	50(98.04%)	51(100.00%)	51(100.00%)	51(100.00%)	50(98.04%)	253(99.22%)
4	50(98.04%)	50(98.04%)	51(100.00%)	51(100.00%)	51(100.00%)	253(99.22%)
5	51(100.00%)	51(100.00%)	51(100.00%)	51(100.00%)	50(98.04%)	254(99.61%)
6	51(100.00%)	51(100.00%)	51(100.00%)	50(98.04%)	50(98.04%)	253(99.22%)
7	51(100.00%)	51(100.00%)	51(100.00%)	51(100.00%)	50(98.04%)	254(99.61%)
8	51(100.00%)	50(98.04%)	51(100.00%)	51(100.00%)	51(100.00%)	254(99.61%)
9	51(100.00%)	51(100.00%)	51(100.00%)	51(100.00%)	51(100.00%)	255(100.00%)
10	51(100.00%)	50(98.04%)	51(100.00%)	51(100.00%)	51(100.00%)	254(99.61%)
Average						253.9 (99.57%)

(F = Fold, R = Run).

Table 6. Offline-Learning Computation Time.

Process	Time (second)
DTCWT	8.41
VE	1.81
TSVM Training	0.29

Table 7. Online-Prediction Computation Time.

Process	Time (second)
DTCWT	0.037
VE	0.009
TSVM Test	0.003

5.6. Comparison to Human Reported Results

In the final experiment, we invited three senior neuroradiologists who have over ten years of experiences. We scanned 20 subjects (5 healthy and 15 pathological), and we pick five slices from each subject. For the healthy subjects, the five slices were selected randomly. For the pathological subjects, the five slices should contain the lesions by confirmation of all the three senior neuroradiologists.

Afterwards, a double blind test was performed. Four junior neuroradiologists with less than 1 year of experiences were required to predict the status of the brain (either pathological or healthy). Each image was assigned with three minutes. Their diagnosis accuracies were listed in Table 8.

Table 8. Comparison to human reported results.

Neuroradiologist	Accuracy
O1	74%
O2	78%
O3	77%
O4	79%
Our method	96%

(O = Observer).

From Table 8, we see our computer-aided diagnosis method achieves an accuracy of 96%. Compared to Table 4, the performance of our proposed method is affected in real-world scenario.

The reasons are complicated. First, the source dataset is downloaded from Harvard medical school, which is for teaching. Hence, the source dataset itself highlighted the difference between pathological and healthy brains intentionally, and the slice positions were selected with care. Second, images from real hospitals are of poorer quality and of poorer localization of lesions. All these contribute to the worsen performance of our method. After all, an accuracy of 96% in rea-world scenario is good and promising.

Another finding is Table 8 is that the four junior neuroradiologists obtained accuracies of 74%, 78%, 77%, and 79%, respectively. All are below 80%. This validates the power of computer vision and machine learning, since computers have proven to deliver better performance in face recognition, video security, *etc.* Nevertheless, there are other more visible symptoms suggesting that something may be wrong in the brain for neuroradiologists. Therefore, this simple test does not reflect the realistic diagnosis accuracy in real hospitals.

6. Conclusions and Future Research

We proposed a novel CAD system for pathological brain detection (PBD) using DTCWT, VE, and TSVM. The results show that the proposed method yields better results than 12 existing methods in terms of classification accuracy.

Our contributions include three points: (1) We investigate the potential use of dual-tree complex wavelet transform (DTCWT) in MR image classification, and prove DTCWT is effective; (2) We utilize twin support vector machine (TSVM) and prove it is better than canonical SVM and GEPSVM; (3) The proposed system "DTCWT + VE + TSVM" is superior to nineteen state-of-the-art systems.

The limitation of our method is the dataset size is too small. We will try to re-check our methods by creating a large dataset. Another limitation is the dataset cannot reflect real-word scenario, thus we need to obtain more data from hospitals directly in the future. The third limitation is that our data involves only middle and late stage of diseases; hence, our method performs not so good for MR images with diseases in early stage.

In the future, we will consider to validate our method use real clinical data and use advanced classification methods, such as RBFNN, deep leaning, least-square techniques. Besides, we will try to apply our method to remote-sensing related fields or hearing loss detection [58]. The advanced parameter estimation, case-based reasoning [59], and optimization [60] techniques will be carried out in a thorough way. Fuzzy method [61] may be applied to remove outliers in the dataset. Coarse-graining [62] can help extract more efficient entropy that is robust to the noises. Video-on-demand [63,64] services may be applied to help reduce computation resources. Particularly, we shall acquire more datasets and compare our method with human interpretation.

Acknowledgments: This paper was supported by Natural Science Foundation of Jiangsu Province (BK20150983), Open Fund of Key laboratory of symbolic computation and knowledge engineering of ministry of education, Jilin University (93K172016K17), Open Fund of Key Laboratory of Statistical information technology and data mining, State Statistics Bureau, (SDL201608), Priority Academic Program Development of Jiangsu Higher Education Institutions (PAPD), Nanjing Normal University Research Foundation for Talented Scholars (2013119XGQ0061, 2014119XGQ0080), Open Fund of Guangxi Key Laboratory of Manufacturing System & Advanced Manufacturing Technology (15-140-30-008K), Open Project Program of the State Key Lab of CAD & CG, Zhejiang University (A1616), Fundamental Research Funds for the Central Universities (LGYB201604).

Author Contributions: Shuihua Wang & Yudong Zhang conceived the study. Yudong Zhang designed the model. Ming Yang acquired the data. Siyuan Lu & Jiquan Yang analyzed the data. Zhengchao Dong interpreted the data. Shuihua Wang developed the program. Yudong Zhang wrote the draft. All authors gave critical revisions and approved the submission.

Conflicts of Interest: The authors declare no conflict of interest.

Abbreviations

The following abbreviations are used in this manuscript:

(A)(BP)(F)(PC)NN	(Artificial) (Back-propagation) (Feed-forward) (Pulse-coupled) neural network
(B)PSO(-MT)	(Binary) Particle Swarm Optimization (-Mutation and TVAC)
(D)(S)W(P)T	(Discrete) (Stationary) wavelet (packet) transform
(k)(F)(LS)(GEP)SVM	(kernel) (Fuzzy) (Least-Squares) (Generalized eigenvalue proximal) Support vector machine
(W)(P)(T)E	(Wavelet) (Packet) (Tsallis) entropy
CAD	Computer-aided diagnosis
CS	Cost-sensitivity
FRFE	Fractional Fourier entropy
HMI	Hu moment invariant

KNN	*K*-nearest neighbors
MR(I)	Magnetic resonance (imaging)
PCA	Principal Component Analysis
RBF	Radial Basis Function
TVAC	Time-varying Acceleration Coefficients
WTT	Welch's *t*-test

References

1. Thorsen, F.; Fite, B.; Mahakian, L.M.; Seo, J.W.; Qin, S.P.; Harrison, V.; Johnson, S.; Ingham, E.; Caskey, C.; Sundstrom, T.; *et al.* Multimodal imaging enables early detection and characterization of changes in tumor permeability of brain metastases. *J. Controll. Release* **2013**, *172*, 812–822.

2. Gorji, H.T.; Haddadnia, J. A novel method for early diagnosis of Alzheimer's disease based on pseudo Zernike moment from structural MRI. *Neuroscience* **2015**, *305*, 361–371.

3. Goh, S.; Dong, Z.; Zhang, Y.; DiMauro, S.; Peterson, B.S. Mitochondrial dysfunction as a neurobiological subtype of autism spectrum disorder: Evidence from brain imaging. *JAMA Psychiatry* **2014**, *71*, 665–671.

4. El-Dahshan, E.S.A.; Hosny, T.; Salem, A.B.M. Hybrid intelligent techniques for MRI brain images classification. *Digit. Signal Process.* **2010**, *20*, 433–441.

5. Patnaik, L.M.; Chaplot, S.; Jagannathan, N.R. Classification of magnetic resonance brain images using wavelets as input to support vector machine and neural network. *Biomed. Signal Process. Control* **2006**, *1*, 86–92.

6. Dong, Z.; Wu, L.; Wang, S.; Zhang, Y. A hybrid method for MRI brain image classification. *Expert Syst. Appl.* **2011**, *38*, 10049–10053.

7. Wu, L. An MR brain images classifier via principal component analysis and kernel support vector machine. *Prog. Electromagn. Res.* **2012**, *130*, 369–388.

8. Das, S.; Chowdhury, M.; Kundu, M.K. Brain MR image classification using multiscale geometric analysis of Ripplet. *Progress Electromagn. Res.-Pier* **2013**, *137*, 1–17.

9. El-Dahshan, E.S.A.; Mohsen, H.M.; Revett, K.; Salem, A.B.M. Computer-Aided diagnosis of human brain tumor through MRI: A survey and a new algorithm. *Expert Syst. Appl.* **2014**, *41*, 5526–5545.

10. Dong, Z.; Ji, G.; Yang, J. Preclinical diagnosis of magnetic resonance (MR) brain images via discrete wavelet packet transform with Tsallis entropy and generalized eigenvalue proximal support vector machine (GEPSVM). *Entropy* **2015**, *17*, 1795–1813.

11. Wang, S.; Dong, Z.; Du, S.; Ji, G.; Yan, J.; Yang, J.; Wang, Q.; Feng, C.; Phillips, P. Feed-Forward neural network optimized by hybridization of PSO and ABC for abnormal brain detection. *Int. J. Imaging Syst. Technol.* **2015**, *25*, 153–164.

12. Nazir, M.; Wahid, F.; Khan, S.A. A simple and intelligent approach for brain MRI classification. *J. Intell. Fuzzy Syst.* **2015**, *28*, 1127–1135.

13. Sun, P.; Wang, S.; Phillips, P.; Zhang, Y. Pathological brain detection based on wavelet entropy and Hu moment invariants. *Bio-Med. Mater. Eng.* **2015**, *26*, 1283–1290.

14. Mount, N.J.; Abrahart, R.J.; Dawson, C.W.; Ab Ghani, N. The need for operational reasoning in data-driven rating curve prediction of suspended sediment. *Hydrol. Process.* **2012**, *26*, 3982–4000.

15. Abrahart, R.J.; Dawson, C.W.; See, L.M.; Mount, N.J.; Shamseldin, A.Y. Discussion of "Evapotranspiration modelling using support vector machines". *Hydrol. Sci. J.-J. Sci. Hydrol.* **2010**, *55*, 1442–1450.

16. Si, Y.; Zhang, Z.S.; Cheng, W.; Yuan, F.C. State detection of explosive welding structure by dual-tree complex wavelet transform based permutation entropy. *Steel Compos. Struct.* **2015**, *19*, 569–583.

17. Hamidi, H.; Amirani, M.C.; Arashloo, S.R. Local selected features of dual-tree complex wavelet transform for single sample face recognition. *IET Image Process.* **2015**, *9*, 716–723.

18. Murugesan, S.; Tay, D.B.H.; Cooke, I.; Faou, P. Application of dual tree complex wavelet transform in tandem mass spectrometry. *Comput. Biol. Med.* **2015**, *63*, 36–41.

19. Smaldino, P.E. Measures of individual uncertainty for ecological models: Variance and entropy. *Ecol. Model.* **2013**, *254*, 50–53.

20. Yang, G.; Zhang, Y.; Yang, J.; Ji, G.; Dong, Z.; Wang, S.; Feng, C.; Wang, Q. Automated classification of brain images using wavelet-energy and biogeography-based optimization. *Multimed. Tools Appl.* **2015**, 1–17.

21. Guang-Shuai, Z.; Qiong, W.; Chunmei, F.; Elizabeth, L.; Genlin, J.; Shuihua, W.; Yudong, Z.; Jie, Y. Automated Classification of Brain MR Images using Wavelet-Energy and Support Vector Machines. In Proceedings of the 2015 International Conference on Mechatronics, Electronic, Industrial and Control Engineering, Shenyang, China, 24–26 April 2015; Liu, C., Chang, G., Luo, Z., Eds.; Atlantis Press: Shenyang, China, 2015; pp. 683–686.

22. Carrasco, M.; Lopez, J.; Maldonado, S. A second-order cone programming formulation for nonparallel hyperplane support vector machine. *Expert Syst. Appl.* **2016**, *54*, 95–104.

23. Wei, Y.C.; Watada, J.; Pedrycz, W. Design of a qualitative classification model through fuzzy support vector machine with type-2 fuzzy expected regression classifier preset. *IEEJ Trans. Electr. Electron. Eng.* **2016**, *11*, 348–356.

24. Wu, L.; Zhang, Y. Classification of fruits using computer vision and a multiclass support vector machine. *Sensors* **2012**, *12*, 12489–12505.

25. Johnson, K.A.; Becker, J.A. The Whole Brain Atlas. Available online: http://www.med.harvard.edu/AANLIB/home.html (accessed on 1 March 2016).

26. Silver, D.; Huang, A.; Maddison, C.J.; Guez, A.; Sifre, L.; van den Driessche, G.; Schrittwieser, J.; Antonoglou, I.; Panneershelvam, V.; Lanctot, M.; *et al.* Mastering the game of Go with deep neural networks and tree search. *Nature* **2016**, *529*, 484–489.

27. Ng, E.Y.K.; Borovetz, H.S.; Soudah, E.; Sun, Z.H. Numerical Methods and Applications in Biomechanical Modeling. *Comput. Math. Methods Med.* **2013**, *2013*, 727830:1–727830:2.

28. Zhang, Y.; Peng, B.; Liang, Y.-X.; Yang, J.; So, K.; Yuan, T.-F. Image processing methods to elucidate spatial characteristics of retinal microglia after optic nerve transection. *Sci. Rep.* **2016**, *6*, 21816.

29. Shin, D.K.; Moon, Y.S. Super-Resolution image reconstruction using wavelet based patch and discrete wavelet transform. *J. Signal. Process. Syst. Signal Image Video Technol.* **2015**, *81*, 71–81.

30. Yu, D.; Shui, H.; Gen, L.; Zheng, C. Exponential wavelet iterative shrinkage thresholding algorithm with random shift for compressed sensing magnetic resonance imaging. *IEEJ Trans. Electr. Electron. Eng.* **2015**, *10*, 116–117.

31. Beura, S.; Majhi, B.; Dash, R. Mammogram classification using two dimensional discrete wavelet transform and gray-level co-occurrence matrix for detection of breast cancer. *Neurocomputing* **2015**, *154*, 1–14.

32. Ayatollahi, F.; Raie, A.A.; Hajati, F. Expression-Invariant face recognition using depth and intensity dual-tree complex wavelet transform features. *J. Electron. Imaging* **2015**, *24*, 13.

33. Hill, P.R.; Anantrasirichai, N.; Achim, A.; Al-Mualla, M.E.; Bull, D.R. Undecimated Dual-Tree Complex Wavelet Transforms. *Signal Process-Image Commun.* **2015**, *35*, 61–70.

34. Kadiri, M.; Djebbouri, M.; Carré, P. Magnitude-Phase of the dual-tree quaternionic wavelet transform for multispectral satellite image denoising. *EURASIP J. Image Video Process.* **2014**, *2014*, 1–16.

35. Singh, H.; Kaur, L.; Singh, K. Fractional M-band dual tree complex wavelet transform for digital watermarking. *Sadhana-Acad. Proc. Eng. Sci.* **2014**, *39*, 345–361.

36. Celik, T.; Tjahjadi, T. Multiscale texture classification using dual-tree complex wavelet transform. *Pattern Recognit. Lett.* **2009**, *30*, 331–339.

37. Zhang, Y.; Wang, S.; Dong, Z.; Phillips, P.; Ji, G.; Yang, J. Pathological brain detection in magnetic resonance imaging scanning by wavelet entropy and hybridization of biogeography-based optimization and particle swarm optimization. *Progress Electromagn. Res.* **2015**, *152*, 41–58.

38. Mangasarian, O.L.; Wild, E.W. Multisurface proximal support vector machine classification via generalized eigenvalues. *IEEE Trans. Pattern Anal. Mach. Intell.* **2006**, *28*, 69–74.

39. Khemchandani, R.; Karpatne, A.; Chandra, S. Generalized eigenvalue proximal support vector regressor. *Expert Syst. Appl.* **2011**, *38*, 13136–13142.

40. Shao, Y.H.; Deng, N.Y.; Chen, W.J.; Wang, Z. Improved Generalized Eigenvalue Proximal Support Vector Machine. *IEEE Signal Process. Lett.* **2013**, *20*, 213–216.

41. Jayadeva; Khemchandani, R.; Chandra, S. Twin support vector machines for pattern classification. *IEEE Trans. Pattern Anal. Mach. Intell.* **2007**, *29*, 905–910.

42. Nasiri, J.A.; Charkari, N.M.; Mozafari, K. Energy-Based model of least squares twin Support Vector Machines for human action recognition. *Signal Process.* **2014**, *104*, 248–257.

43. Xu, Z.J.; Qi, Z.Q.; Zhang, J.Q. Learning with positive and unlabeled examples using biased twin support vector machine. *Neural Comput. Appl.* **2014**, *25*, 1303–1311.

44. Shao, Y.H.; Chen, W.J.; Zhang, J.J.; Wang, Z.; Deng, N.Y. An efficient weighted Lagrangian twin support vector machine for imbalanced data classification. *Pattern Recognit.* **2014**, *47*, 3158–3167.

45. Zhang, Y.-D.; Wang, S.-H.; Yang, X.-J.; Dong, Z.-C.; Liu, G.; Phillips, P.; Yuan, T.-F. Pathological brain detection in MRI scanning by wavelet packet Tsallis entropy and fuzzy support vector machine. *SpringerPlus* **2015**, *4*, 716.

46. Zhang, Y.-D.; Chen, S.; Wang, S.-H.; Yang, J.-F.; Phillips, P. Magnetic resonance brain image classification based on weighted-type fractional Fourier transform and nonparallel support vector machine. *Int. J. Imaging Syst. Technol.* **2015**, *25*, 317–327.

47. Zhang, Y.; Wang, S. Detection of Alzheimer's disease by displacement field and machine learning. *PeerJ* **2015**, *3*.

48. Purushotham, S.; Tripathy, B.K. Evaluation of classifier models using stratified tenfold cross validation techniques. In *Global Trends in Information Systems and Software Applications*; Krishna, P.V., Babu, M.R., Ariwa, E., Eds.; Springer-Verlag Berlin: Berlin, Germany, 2012; Volume 270, pp. 680–690.

49. Ng, E.Y.K.; Jamil, M. Parametric sensitivity analysis of radiofrequency ablation with efficient experimental design. *Int. J. Thermal Sci.* **2014**, *80*, 41–47.

50. Kumar, M.A.; Gopal, M. Least squares twin support vector machines for pattern classification. *Expert Syst. Appl.* **2009**, *36*, 7535–7543.

51. Zhuang, J.; Widschwendter, M.; Teschendorff, A.E. A comparison of feature selection and classification methods in DNA methylation studies using the Illumina Infinium platform. *BMC Bioinform.* **2012**, *13*, 14.

52. Shamsinejadbabki, P.; Saraee, M. A new unsupervised feature selection method for text clustering based on genetic algorithms. *J. Intell. Inf. Syst.* **2012**, *38*, 669–684.

53. Dong, Z.; Liu, A.; Wang, S.; Ji, G.; Zhang, Z.; Yang, J. Magnetic resonance brain image classification via stationary wavelet transform and generalized eigenvalue proximal support vector machine. *J. Med. Imaging Health Inform.* **2015**, *5*, 1395–1403.

54. Yang, X.; Sun, P.; Dong, Z.; Liu, A.; Yuan, T.-F. Pathological Brain Detection by a Novel Image Feature—Fractional Fourier Entropy. *Entropy* **2015**, *17*, 7877.

55. Zhou, X.-X.; Yang, J.-F.; Sheng, H.; Wei, L.; Yan, J.; Sun, P.; Wang, S.-H. Combination of stationary wavelet transform and kernel support vector machines for pathological brain detection. *Simulation* **2016**.

56. Sun, N.; Morris, J.G.; Xu, J.; Zhu, X.; Xie, M. iCARE: A framework for big data-based banking customer analytics. *IBM J. Res. Dev.* **2014**, *58*, 9.

57. Rajeshwari, J.; Karibasappa, K.; Gopalakrishna, M.T. Three phase security system for vehicles using face recognition on distributed systems. In *Information Systems Design and Intelligent Applications*; Satapathy, S.C., Mandal, J.K., Udgata, S.K., Bhateja, V., Eds.; Springer-Verlag Berlin: Berlin, Germany, 2016; Volume 435, pp. 563–571.

58. Wang, S.; Yang, M.; Zhang, Y.; Li, J.; Zou, L.; Lu, S.; Liu, B.; Yang, J.; Zhang, Y. Detection of Left-Sided and Right-Sided Hearing Loss via Fractional Fourier Transform. *Entropy* **2016**, *18*, 194.

59. Shubati, A.; Dawson, C.W.; Dawson, R. Artefact generation in second life with case-based reasoning. *Softw. Qual. J.* **2011**, *19*, 431–446.

60. Zhang, Y.; Yang, X.; Cattani, C.; Rao, R.; Wang, S.; Phillips, P. Tea Category Identification Using a Novel Fractional Fourier Entropy and Jaya Algorithm. *Entropy* **2016**, *18*, 77.

61. Ji, L.Z.; Li, P.; Li, K.; Wang, X.P.; Liu, C.C. Analysis of short-term heart rate and diastolic period variability using a refined fuzzy entropy method. *Biomed. Eng. Online* **2015**, *14*, 13.

62. Li, P.; Liu, C.Y.; Li, K.; Zheng, D.C.; Liu, C.C.; Hou, Y.L. Assessing the complexity of short-term heartbeat interval series by distribution entropy. *Med. Biol. Eng. Comput.* **2015**, *53*, 77–87.

63. Lau, P.Y.; Park, S. A new framework for managing video-on-demand servers: Quad-Tier hybrid architecture. *IEICE Electron. Express* **2011**, *8*, 1399–1405.

64. Lau, P.Y.; Park, S.; Lee, J. Cohort-Surrogate-Associate: A server-subscriber load sharing model for video-on-demand services. *Malayas. J. Comput. Sci.* **2011**, *24*, 1–16.

2D Gaze Estimation Based on Pupil-Glint Vector Using an Artificial Neural Network

Jianzhong Wang, Guangyue Zhang and Jiadong Shi

Abstract: Gaze estimation methods play an important role in a gaze tracking system. A novel 2D gaze estimation method based on the pupil-glint vector is proposed in this paper. First, the circular ring rays location (CRRL) method and Gaussian fitting are utilized for pupil and glint detection, respectively. Then the pupil-glint vector is calculated through subtraction of pupil and glint center fitting. Second, a mapping function is established according to the corresponding relationship between pupil-glint vectors and actual gaze calibration points. In order to solve the mapping function, an improved artificial neural network (DLSR-ANN) based on direct least squares regression is proposed. When the mapping function is determined, gaze estimation can be actualized through calculating gaze point coordinates. Finally, error compensation is implemented to further enhance accuracy of gaze estimation. The proposed method can achieve a corresponding accuracy of 1.29°, 0.89°, 0.52°, and 0.39° when a model with four, six, nine, or 16 calibration markers is utilized for calibration, respectively. Considering error compensation, gaze estimation accuracy can reach 0.36°. The experimental results show that gaze estimation accuracy of the proposed method in this paper is better than that of linear regression (direct least squares regression) and nonlinear regression (generic artificial neural network). The proposed method contributes to enhancing the total accuracy of a gaze tracking system.

Reprinted from *Appl. Sci.* Cite as: Wang, J.; Zhang, G.; Shi, J. 2D Gaze Estimation Based on Pupil-Glint Vector Using an Artificial Neural Network. *Appl. Sci.* **2016**, *6*, 174.

1. Introduction

Human beings acquire 80%–90% of outside information through the eyes. Humans' visual perception of information can be acquired through eye gaze tracking [1–4]. With the increasing development of computer/machine vision technology, gaze tracking technology has been more and more widely applied in the fields of medicine [5], production tests [6], human–machine interaction [7,8], military aviation [9,10], *etc.*

According to differences in dimension of gaze direction estimation, gaze tracking technology can be divided into 2D gaze tracking [11–19] and 3D gaze tracking [20–27]; according to differences in ways of wearing, gaze tracking technology can be classed as intrusive (head-mounted) [12,28–37] or non-intrusive (head-free) [20,23,38–44].

For different gaze tracking systems, gaze tracking methods mainly contain Limbus Tracking [45–47], Pupil Tracking [48–50], Pupil-glint Vector [51–55], Purkinje Image [24,56,57], *etc.*

For 2D gaze estimation methods, mapping function between gaze points and target plane or regions of interest is firstly established. The mapping function solved is then further utilized to calculate the gaze point on certain targets or regions. For 3D gaze estimation methods, a human eyeball model is employed to determine the absolute position of eyes in the test space. On this basis, 3D gaze is calculated to acquire the specific staring location or fixation targets of human eyes in the space.

The main purpose of this paper is to estimate the gaze point of the human eye on a monitor screen fixed to the head. The mapping function between gaze points and fixation targets is from plane to plane. Therefore, a novel 2D gaze estimation method based on pupil-glint vector is proposed to calculate the gaze direction.

In conventional 2D gaze estimation methods, the most widely utilized calculation methods can be divided into two groups: linear regression (direct least squares regression) [58–64] and nonlinear regression (generic artificial neural network) [65–69]. In [58–61], Morimoto *et al.* utilize least squares to calculate the mapping function between calibration markers and corresponding pupil-glint vectors. Overdetermined linear equations for solving mapping function are composed by a series of 2nd-order polynomials. The number of polynomials depends on the number of calibration markers. Corresponding coordinates of calibration markers, pupil and glint centers are determined through a calibration process. The pupil-glint vector is calculated through the subtraction of pupil and glint center coordinates. Cherif *et al.* [62] propose an adaptive calibration method. A second time calibration is employed for error correction. A polynomial transformation of higher order is utilized to model mapping function by applying a mean square error criterion. The result of single calibration shows that the gaze estimation accuracy will increase with the enhancement of the polynomial order. However, through experimental analyses, Cerrolaza *et al.* [63,64] point out that the gaze estimation accuracy of a gaze tracking system will not increase with the enhancement of polynomial order owing to the factors of head motion, number of calibration markers, and calculating method of pupil-glint vector, *etc.* When applied for solving mapping function, 2nd-order linear polynomial is the most widely used linear regression solution method with the advantages of fewer calibration markers and better approximation effect.

An artificial neural network is the most widely used nonlinear regression method for solving the mapping function between calibration markers and corresponding pupil-glint vectors (or pupil centers, eye movements information, *etc.*). Early in 1994, Baluja and Pomerleau [65] proposed the method using simple artificial neural network (ANN) to estimate gaze direction. Multi-group attempts are conducted to find a training network with optimal performance. In the first attempt, images of only the pupil and cornea are utilized as inputs to ANN. The output units are organized with horizontal and vertical coordinates of the gaze point. A single divided layer is used for training in the ANN architecture. In the second attempt, in order to achieve a better accuracy, the total eye socket (including pupil and glint position) is utilized as an input to ANN. A single continuous hidden layer and a single divided hidden layer are used for training in the ANN architecture. The experimental results show that when the hidden layer units are fewer, the training accuracy of the divided hidden layer is higher than that of the continuous hidden layer. In addition, the training time is short. Furthermore, some of the eye images are employed as training sets and the remainder are employed as testing sets, which provides more accurate experimental results. However, though a higher accuracy can be achieved when the total eye socket (including pupil and glint position) is utilized as an input to ANN, the training sample data is huge and the training time is long. Piratla *et al.* [66] developed a network-based gaze tracking system. As an auxiliary tool, a strip with black and white bands is mounted on the user's head to facilitate real-time eye detection. Twelve items, consisting of strip edge coordinates at lower ends, eyeball centers coordinates, and eyelid distances, are the input features of the neural network. The X and Y coordinate pair of the point the user is looking at on the screen is the output of the neural network. A 25-neuron hidden layer is utilized between the input and output layer. This method requires a large number of input items and a long detection period. The real-time quality needs to be improved. Demjen *et al.* [67] compare the neural network and linear regression method utilized for estimating gaze direction. The comparison results show that: (1) the calibration procedure of the neural network method is faster as it requires fewer calibration markers, and (2) the neural network method provides higher accuracy. The gaze tracking performance of a neural network is better than that of linear regression. Multi-layer perceptrons (MLPs) are utilized by Coughlin *et al.* [68] to calculate gaze point coordinates based on electro-oculogram (EOG). The number of input nodes depends on the number of data points chosen to represent the saccadic waveforms. The output nodes of the network provide the horizontal and vertical 2D spatial coordinates of the line-of-sight on a particular training or test trial. In order to determine the number of nodes that can provide the optimal outputs, hidden layers containing different numbers of nodes are selected to train MLP ANN. Initial weights trained on another person are referred to in order to reduce training time. The experimental results show that using MLPs for calibration appears to be able to overcome some of the disadvantages of the EOG and provides an accuracy not significantly different from that obtained with the infrared tracker. In addition, Sesin *et al.* [69] find that MLPs can produce positive effectives: jitter reduction of gaze point estimation and enhancing the calculating stability of gaze points. Gneo *et al.* [70] utilize multilayer neural feedforward networks (MFNNs) to calculate gaze point coordinates based on pupil-glint vectors. Two separate MFNNs (each one having the same eye features as inputs, with one single output neuron directly estimating one of the X and Y coordinates of the POG), each containing 10 neurons in the hidden layer, are employed for training to acquire the outputs. The use of MFNNs overcomes the drawbacks of the model-based EGTSs and the potential reasons for their failure, which sometimes give ANNs an undeservedly poor reputation. Zhu and Ji [71] utilize generalized regression neural networks (GRNNs) to calculate a mapping function from pupil parameters to screen coordinates in a calibration procedure. The GRNN topology consists of four layers: input layer, hidden layer, summation layer, and output layer. Six factors including pupil-glint vector, pupil ellipse orientation, *etc.* are chosen as the input parameters of GRNNs. The output nodes represent the horizontal and vertical coordinates of the gaze point. Though the use of hierarchical classification schemes simplifies the calibration procedure, the gaze estimation accuracy of this method is not perfect. Kiat and Ranganath [72] utilize two single radial basis function neural networks (RBFNNs) to map the complex and non-linear relationship between the pupil and glint parameters (inputs) to the gaze point on the screen (outputs). Both of the networks have 11 inputs including x and y coordinates of left and right pupils, pupil-to-glint vectors of the left and right eyes, *etc.* The number of network output nodes depends on the number of calibration regions in the horizontal and vertical direction. The weights of the network are stored as calibration data for every subsequent time the user operates the system. As is the case with GRNNs, the gaze estimation accuracy of RBFNNs is not high enough. Wu *et al.* [73] employ the Active Appearance Model (AAM) to represent the eye image features, which combines the shape and texture information in the eye region. The support vector machine (SVM) is utilized to classify 36 2D eye feature points set (including eye contour, iris and pupil parameters, *etc.*) into eye gazing direction. The final results show the independence of the classifications and the accurate estimation of the gazing directions.

In this paper, considering the high speed of direct least squares regression and the high accuracy of artificial neural network, we propose an improved artificial neural network based on direct least squares regression (DLSR-ANN) to calculate the mapping function between pupil-glint vectors and actual gaze points. Different from general artificial neural networks, coefficient matrix elements of direct least squares regression are employed as connection coefficients in the input and hidden layers of DLSR-ANN. The error cost function and continuous-time learning rule of DLSR-ANN are defined and calculated according to the constraint condition of solving direct least squares regression. The initial condition of an integrator associated with the learning rule of DLSR-ANN is acquired through linear polynomial calculation of direct least squares regression. The learning rate parameter is limited to a range determined by the maximal eigenvalue of auto-correlation matrix composed by input vector of direct least squares regression. The proposed method contains advantages of both direct least squares regression and artificial neural network.

The remainder of this paper is organized as follows: Section 2 presents the proposed neural network method for gaze estimation in detail. Section 3 describes the experimental system and shows the results. Section 4 concludes the whole work. The experimental results show that the training process of the proposed method is stable. The gaze estimation accuracy of the proposed method in this paper is better than that of conventional linear regression (direct least squares regression) and nonlinear regression (generic artificial neural network). The proposed method contributes to enhance the total accuracy of a gaze tracking system.

2. Proposed Methods for Gaze Estimation

According to the respective characteristics of linear and nonlinear regression, a novel 2D gaze estimation method based on pupil-glint vector is proposed in this paper. An improved artificial neural network (DLSR-ANN) based on direct least squares regression is developed to solve the mapping function between pupil-glint vector and gaze point and then calculate gaze direction. The flow-process of gaze direction estimation is shown in Figure 1. First, when gazing at the calibration markers on the screen, corresponding eye images of subjects are acquired through a camera fixed on the head-mounted gaze tracking system. Second, through preprocessing such as Otsu optimal threshold binarization and opening-and-closing operation, pupil and glint centers are detected by utilizing circular ring rays location (CRRL) method. As inputs of the proposed DLSR-ANN, pupil-glint vector is calculated through the subtraction of pupil and glint center coordinates. Third, a three-layer DLSR-ANN (input layer, hidden layer, and output layer) is developed to calculate the mapping function between pupil-glint vectors and corresponding gaze points. Finally, gaze points on the screen can be estimated according to the mapping function determined.

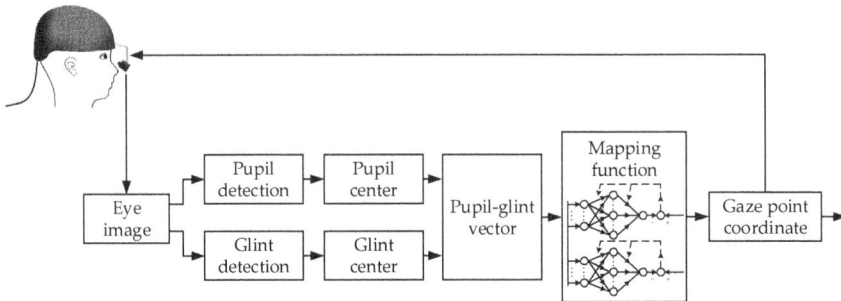

Figure 1. Flow-process of gaze direction estimation.

Pupil-glint vector is calculated through the subtraction of pupil and glint center coordinate. 2nd linear gaze mapping function based on pupil-glint vector is expressed as Equation (1).

$$\begin{cases} x_{ci} = a_1 + a_2 x_{ei} + a_3 y_{ei} + a_4 x_{ei} y_{ei} + a_5 x_{ei}^2 + a_6 y_{ei}^2 \\ y_{ci} = b_1 + b_2 x_{ei} + b_3 y_{ei} + b_4 x_{ei} y_{ei} + b_5 x_{ei}^2 + b_6 y_{ei}^2 \end{cases} \tag{1}$$

where $i = 1, 2, \cdots, N$. N is the number of calibration markers. (x_{ci}, y_{ci}) is the coordinate of gaze calibration markers on screen coordinate system. (x_{ei}, y_{ei}) is the coordinate of pupil-glint vector on image coordinate system. Least squares, as conventional linear methods, is utilized to solve the gaze mapping function shown in Equation (1). Residual error is defined as:

$$R^2 = \sum_{i=1}^{N} \left[x_{ci} - \left(a_1 + a_2 x_{ei} + a_3 y_{ei} + a_4 x_{ei} y_{ei} + a_5 x_{ei}^2 + a_6 y_{ei}^2 \right) \right]^2. \tag{2}$$

By calculating a partial derivative of a_j $(j = 1, 2, 3, 4, 5, 6)$ in Equation (2), the constraint condition can be obtained as in Equation (3).

$$\frac{\partial R^2}{\partial a_j} = -2 \sum_{i=1}^{N} \sigma_j \left[x_{ci} - \left(a_1 + a_2 x_{ei} + a_3 y_{ei} + a_4 x_{ei} y_{ei} + a_5 x_{ei}^2 + a_6 y_{ei}^2 \right) \right] = 0, \tag{3}$$

where $\sigma_1 = 1$, $\sigma_2 = x_{ei}$, $\sigma_3 = y_{ei}$, $\sigma_4 = x_{ei} y_{ei}$, $\sigma_5 = x_{ei}^2$, $\sigma_6 = y_{ei}^2$. The value of a_j can be calculated according to Equation (4).

$$\begin{bmatrix} \sum_{i=1}^{N} x_{ci} \\ \sum_{i=1}^{N} x_{ci} x_{ei} \\ \sum_{i=1}^{N} x_{ci} y_{ei} \\ \sum_{i=1}^{N} x_{ci} x_{ei} y_{ei} \\ \sum_{i=1}^{N} x_{ci} x_{ei}^2 \\ \sum_{i=1}^{N} x_{ci} y_{ei}^2 \end{bmatrix} = \begin{bmatrix} N & \sum_{i=1}^{N} x_{ei} & \sum_{i=1}^{N} y_{ei} & \sum_{i=1}^{N} x_{ei} y_{ei} & \sum_{i=1}^{N} x_{ei}^2 & \sum_{i=1}^{N} y_{ei}^2 \\ \sum_{i=1}^{N} x_{ei} & \sum_{i=1}^{N} x_{ei}^2 & \sum_{i=1}^{N} x_{ei} y_{ei} & \sum_{i=1}^{N} x_{ei}^2 y_{ei} & \sum_{i=1}^{N} x_{ei}^3 & \sum_{i=1}^{N} x_{ei} y_{ei}^2 \\ \sum_{i=1}^{N} y_{ei} & \sum_{i=1}^{N} x_{ei} y_{ei} & \sum_{i=1}^{N} y_{ei}^2 & \sum_{i=1}^{N} x_{ei} y_{ei}^2 & \sum_{i=1}^{N} x_{ei}^2 y_{ei} & \sum_{i=1}^{N} y_{ei}^3 \\ \sum_{i=1}^{N} x_{ei} y_{ei} & \sum_{i=1}^{N} x_{ei}^2 y_{ei} & \sum_{i=1}^{N} x_{ei} y_{ei}^2 & \sum_{i=1}^{N} x_{ei}^2 y_{ei}^2 & \sum_{i=1}^{N} x_{ei}^3 y_{ei} & \sum_{i=1}^{N} x_{ei} y_{ei}^3 \\ \sum_{i=1}^{N} x_{ei}^2 & \sum_{i=1}^{N} x_{ei}^3 & \sum_{i=1}^{N} x_{ei}^2 y_{ei} & \sum_{i=1}^{N} x_{ei}^3 y_{ei} & \sum_{i=1}^{N} x_{ei}^4 & \sum_{i=1}^{N} x_{ei}^2 y_{ei}^2 \\ \sum_{i=1}^{N} y_{ei}^2 & \sum_{i=1}^{N} x_{ei} y_{ei}^2 & \sum_{i=1}^{N} y_{ei}^3 & \sum_{i=1}^{N} x_{ei} y_{ei}^3 & \sum_{i=1}^{N} x_{ei}^2 y_{ei}^2 & \sum_{i=1}^{N} y_{ei}^4 \end{bmatrix} \begin{bmatrix} a_1 \\ a_2 \\ a_3 \\ a_4 \\ a_5 \\ a_6 \end{bmatrix} \tag{4}$$

As with a_j, the value of b_j $(j = 1, 2, 3, 4, 5, 6)$ can be calculated. In fact, the relationship between the number of coefficients in mapping function (r) and polynomial order (s) is as follows:

$$r = 1 + \sum_{t=1}^{s} (t + 1). \tag{5}$$

According to Equation (5), when an s order polynomial is utilized to solve the gaze mapping function, at least r gaze calibration markers are required. For a head-mounted (intrusive) gaze tracking system, the relative position of the monitor screen and the user's head and eyes remains nearly fixed. In this case, the higher-order terms in the mapping function are mainly utilized to compensate for error between the estimated and actual gaze direction. The higher the polynomial order, the higher the calculation accuracy. However, the number of polynomial coefficients to be solved will increase at the same time (Equation (5)). In addition, the number of calibration markers required also increases. This not only makes the calibration time longer; the cumbersome calibration process also adds to the user's burden. Users are prone to be fatigued, thus affecting the calibration accuracy. In order to further enhance the mapping accuracy and realize precise estimation of gaze direction, a novel artificial neural network (DLSR-ANN) based on direct least squares regression is proposed to solve a mapping function between pupil-glint vectors and calibration markers.

We rewrite the matrix equation in Equation (4) as:

$$Qa = p, \tag{6}$$

where $a = \begin{bmatrix} a_1 & a_2 & a_3 & a_4 & a_5 & a_6 \end{bmatrix}^{\mathrm{T}}$, $p = \left[\sum_{i=1}^{N} x_{ci} \quad \sum_{i=1}^{N} x_{ci} x_{ei} \quad \sum_{i=1}^{N} x_{ci} y_{ei} \quad \sum_{i=1}^{N} x_{ci} x_{ei} y_{ei} \right.$ $\left. \sum_{i=1}^{N} x_{ci} x_{ei}^2 \quad \sum_{i=1}^{N} x_{ci} y_{ei}^2 \right]^{\mathrm{T}}$, Q is the coefficient matrix.

Figure 2 shows the scheme framework of an improved artificial neural network based on direct least squares regression. The DLSR-ANN is a three-layer neural network with input layer, hidden layer, and output layer. Elements of matrix p including pupil-glint vectors gazing at calibration markers are determined as the input of a neural network. Elements of matrix a are determined as the output of a neural network. The input, output, and hidden layers contain one, one, and three nodes, respectively.

As shown in Figure 2, coefficient matrix elements of direct least squares regression are employed as connection coefficients in the input and hidden layers of DLSR-ANN. According to the respective characteristics of input, hidden, and output layers and the relationship among them, appropriate weighting functions $g_1(t)$, $g_2(t)$, $g_3(t)$ are determined. Derivatives of $g_1(t)$, $g_2(t)$, and $g_3(t)$, respectively, are calculated $(f_1(t), f_2(t), f_3(t)$ $(f(t) = dg(t)/dt))$ as the transfer function of the neuron. The selection of specific parameters is described in Section 3.4. As a three-layer neural network, its output layer carries an integrator. The integrator's initial condition $a(0) = a_0 = \begin{bmatrix} a_1(0) & a_2(0) & a_3(0) & a_4(0) & a_5(0) & a_6(0) \end{bmatrix}^{\mathrm{T}}$ is calculated through a linear polynomial solution utilizing direct least squares regression.

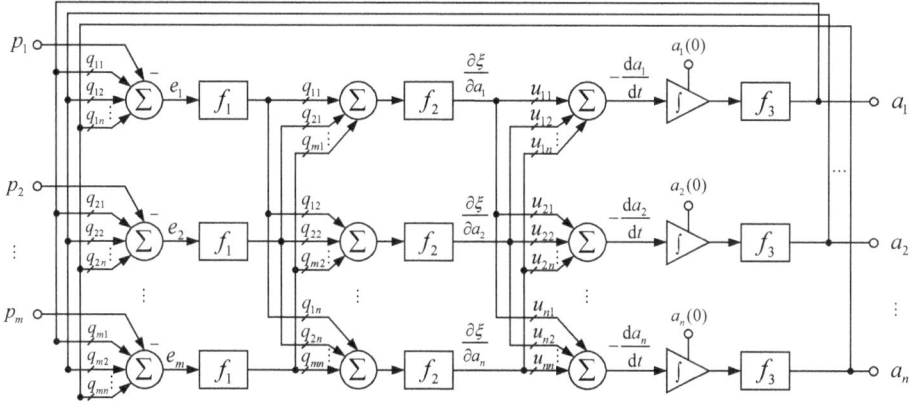

Figure 2. Scheme framework of improved artificial neural network based on direct least squares regression.

In the proposed method, to solve the mapping function in Equation (6), the steepest gradient descent method [74] is adopted as the training method of the neural network. To determine the relationship between hidden layer and output layer, the error cost function and continuous-time learning rule of DLSR-ANN are defined according to the constraint condition of solving direct least squares regression. According to the error distribution characteristics of gaze estimation, the Euclid norm (L_2 norm) is selected to acquire the minimal error cost function, which is in the same form as the error solving criterion of direct least squares regression, as defined in Equation (7):

$$\xi(a) = \frac{1}{2} \| e \|^2 = \frac{1}{2} e^T e, \tag{7}$$

where $e = Qa - p$ is the solution error of Equation (6) in direct least squares regression.

Equation (7) can be further expressed as follows:

$$\begin{aligned} \xi(a) &= \frac{1}{2} (Qa - p)^T (Qa - p) \\ &= \frac{1}{2} \left(a^T Q^T Qa - a^T Qp - p^T Qa + p^T p \right). \end{aligned} \tag{8}$$

According to an error cost function based on the constraint condition of direct least squares regression, the learning rule of a continuous-time neural network is set as Equation (9). The function of the learning rule is to modify the weights of DLSR-ANN adaptively to acquire the optimal solution.

$$\frac{da}{dt} = -\mu \nabla_a \xi(a) = -\mu \frac{\partial \xi(a)}{\partial a}, \tag{9}$$

175

where μ is the learning rate parameter. As a positive-definite matrix, μ ($\mu = [\mu_{vw}]$, $v, w = 1, 2, \cdots, n$) is generally selected as a diagonal matrix. In general, μ is determined by experience. If μ is set too small, the weights of the neural network will be modified by the learning rule slowly. More iterations will be needed to reach the error bottom. If μ is set too large, the learning rule will show numerical instability. To ensure the stability of the differential equation in Equation (9) and the convergence of its solution, a small enough μ is chosen according to Equation (10):

$$0 < \mu < \frac{2}{\lambda_{max}}, \tag{10}$$

where λ_{max} is the maximal eigenvalue of auto-correlation matrix composed by input vector p in direct least squares regression. When the eigenvalue is unavailable, the auto-correlation matrix can replace it.

By calculating a partial derivative of variable a in Equation (8), the learning rule of a continuous-time neural network for solving matrix equation $Qa = p$ can be deduced as:

$$\frac{da}{dt} = -\mu \frac{\partial \xi(a)}{\partial a} = -\mu \cdot \frac{1}{2} \left(2Q^T Qa - 2Q^T p \right) = -\mu \cdot Q^T \left(Qa - 2Q^T p \right) = \mu Q^T e. \tag{11}$$

3. Experimental System and Results

3.1. Experimental System

In this study, we develop a wearable gaze tracking system composed of a helmet, a monitor, an array of four near-infrared light emitting diodes (NIR LEDs), and a microspur camera, as shown in Figure 3. The screen size of the monitor is 75 mm $\times$ 50 mm. Considering the imaging distance is limited between 3 cm and 7 cm, a microspur camera is adopted to acquire the eye image. The image resolution is 640 $\times$ 480 pixels (CCD sensor). As described in [75], when the wavelength of NIR LED is located within the range of 760 nm–1400 nm, the pupil absorbs nearly all the near-infrared light and the iris obviously reflects it. The wavelength of NIR LED employed in this paper is 850 nm and the power is less than 5 mw. The experimental system brings no harm to human eyes [76]. An NVIDIA Jetson TK1 embedded development board (Figure 4) is utilized for image acquiring and processing (NVIDIA: NVIDIA Corporation (Santa Clara, California, CA, USA). TK1: Tegra K1. Jetson TK1 is a code of embedded development board manufactured by NVIDIA Corporation).

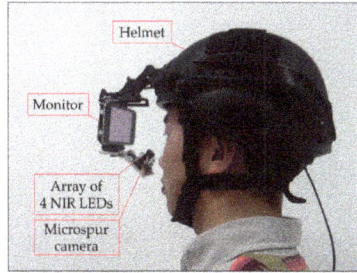

Figure 3. Wearable gaze tracking system.

Figure 4. NVIDIA Jetson TK1 embedded development board.

3.2. Pupil and Glint Detection

3.2.1. Pupil Detection

The circular ring rays location (CRRL) method [77] is utilized for pupil center detection, for the reason that it is more robust and accurate than conventional detection methods. As shown in Figure 5, in the CRRL method, improved Otsu optimal threshold binarization is utilized on a gray-scale eye image to eliminate the influence caused by illumination change. Through an opening-and-closing operation, rough location of pupil area, and circular ring rays, and pupil boundary points and center can be detected accurately when interference factors such as eyelashes, glint, and natural light reflection are located on the pupil contour. The CRRL method contributes to enhance the stability, accuracy, and real-time quality of a gaze tracking system.

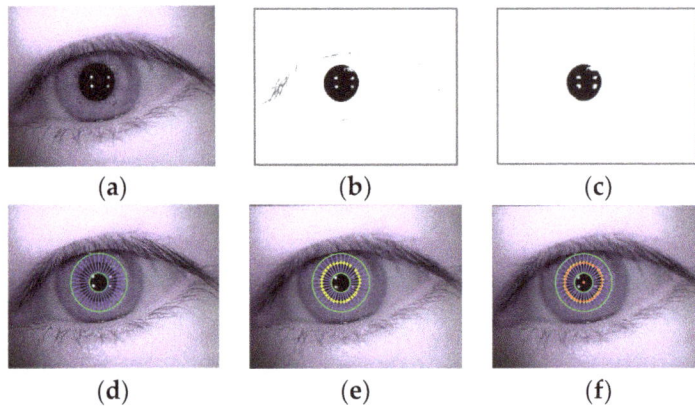

Figure 5. Pupil detection: (**a**) original eye image; (**b**) eye binary image utilizing improved Otsu optimal threshold; (**c**) results of opening-and-closing operation; (**d**) rough location of pupil region; (**e**) extraction of pupil boundary points; (**f**) pupil contour fitting.

3.2.2. Glint Detection

For the reason that the glint's illumination intensity is suitable for Gaussian distribution, Gaussian function deformation solved by improved total least squares [77] is utilized to calculate the glint center. The detection result of glint is shown in Figure 6.

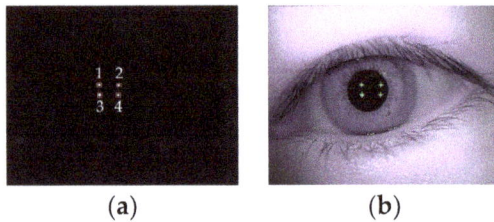

Figure 6. Glint detection: (**a**) rough location of glint; (**b**) glint detection results.

As a sample, some of the pupil and glint centers detected are shown in Table 1.

Table 1. A sample of the pupil and glint centers.

Eye Image	Pupil Center (x, y)	Glint Center (x, y)			
		1	2	3	4
1	(290.15, 265.34)	(265.31, 298.65)	(294.56, 300.87)	(266.41, 310.28)	(296.25, 312.49)
2	(251.42, 255.93)	(245.58, 292.36)	(276.54, 295.13)	(246.19, 305.67)	(277.51, 307.26)
3	(203.34, 260.81)	(221.95, 297.32)	(252.49, 298.61)	(221.34, 309.17)	(253.65, 310.28)
4	(297.74, 275.62)	(271.25, 300.56)	(301.58, 300.67)	(270.91, 315.66)	(300.85, 315.46)
5	(247.31, 277.58)	(243.25, 302.62)	(273.55, 303.46)	(242.81, 317.54)	(274.26, 318.19)

3.3. Calibration Model

As expressed in Equation (5), at least three, six, and 10 polynomial coefficients are required to be calculated, respectively, when a 1st, 2nd, and 3rd order linear polynomial is utilized for calibration, which means that at least three, six, and 10 calibration markers are required. When the number of calibration markers needed is too large, unessential input items can be removed according to principal component analysis to reduce the number of polynomial coefficients to be solved. Generally, based on an overall consideration of the real-time quality and accuracy of a gaze tracking system, four- and five-marker calibration models are most widely employed for 1st order calculation, while six- and nine-marker calibration models are most widely employed for 2nd order calculation [78,79].

Considering that there is some motion between the wearable gaze tracking system and the user's head, error of gaze point data will occur along with a drifting motion. In this paper, position coordinates of quadrangular NIR LEDs are considered as inputs of gaze estimation model to compensate for error caused by drifting motion. As shown in Figure 7, for the purpose of comparison, four-, six-, nine-, and 16-marker calibration models are employed in the process of calculating mapping function. Gaze direction is estimated with and without error compensation. The gaze tracking accuracy of the two cases is compared.

3.4. Gaze Point Estimation

An improved artificial neural network (DLSR-ANN) based on direct least squares regression is developed to calculate the mapping function between pupil-glint vectors and calibration markers. For four-, six-, nine-, and 16-marker calibration models, the number of training samples is selected as 180. The number of hidden nodes is equal to the number of training samples. The x

(or y) coordinate is set as the output of the neural network. Two separate DLSR-ANNs are utilized to estimate the x and y coordinates of the gaze point on the screen. Each separate neural network has the same inputs. Weighting function $g_1(t)$, $g_2(t)$, $g_3(t)$ is respectively determined as $g_1(t) = \frac{1-e^{-\beta_1 t^2}}{1+e^{-\beta_1 t^2}}$, $g_2(t) = \begin{cases} \frac{t^2}{2}, t \leqslant \beta_2 \\ \beta_2 |t| - \frac{\beta_2^2}{2}, t > \beta_2 \end{cases}$, $g_3(t) = \beta_3^2 \ln\left(\cosh\frac{t}{\beta_3}\right)$. The transfer function for input, hidden, and output layers is selected as the derivative of $g_1(t)$, $g_2(t)$, $g_3(t)$, which is respectively calculated as $f_1(t)$, $f_2(t)$, $f_3(t)$: $f_1(t) = \frac{dg_1(t)}{dt} = \frac{4\beta_1 t e^{-\beta_1 t^2}}{\left(1+e^{-\beta_1 t^2}\right)^2}$, $f_2(t) = \frac{dg_2(t)}{dt} = \begin{cases} -\beta_2, t \leqslant -\beta_2 \\ t, |t| \leqslant \beta_2 \\ \beta_2, t > \beta_2 \end{cases}$, $f_3(t) = \frac{dg_3(t)}{dt} = \beta_3 \tanh\frac{t}{\beta_3}$. Learning rate parameter μ is determined by $\mu = \mu_j = 0.0025$ (when a four-marker calibration model is employed, $j = 1, 2, 3, 4$; when a six-, nine-, or 16-marker calibration model is employed, $j = 1, 2, 3, 4, 5, 6$). In order to acquire optimal learning and training results, $\beta_1, \beta_2, \beta_3$ is respectively determined as $\beta_1 = 0.8$, $\beta_2 = 0.7$, $\beta_3 = 0.7$ through a process of trial and error. The initial condition $a(0)$ of an integrator associated with learning rules is acquired through linear polynomial calculation in direct least squares regression.

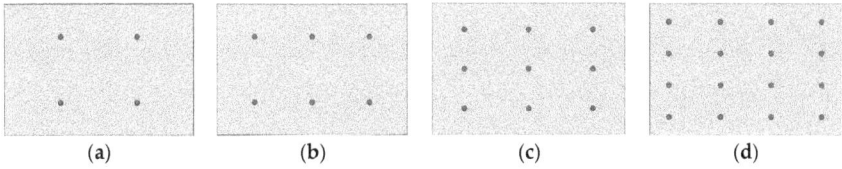

(a)   (b)   (c)   (d)

Figure 7. Calibration model: (**a**) four-marker calibration model; (**b**) six-marker calibration model; (**c**) nine-marker calibration model; (**d**) 16-marker calibration model.

In the developed wearable gaze tracking system, an array of four near-infrared light emitting diodes (NIR LEDs) is employed instead of the conventional single one. The NIR LEDs array can form well-distributed illumination around the human eye, which contributes to extract pupil and glint characteristics more stably and precisely. In addition the center position coordinates of quadrangular NIR LEDs, considered as inputs of the neural network, can further compensate for error caused during the process of gaze point calculation. When a calibration process is accomplished, a model with 8 × 8 test markers is employed to validate the calculation accuracy of the gaze point. Figure 8a–d shows the gaze point estimated through the proposed method with/without considering error compensation, utilizing a four-, six-, nine-, or 16-marker calibration model, respectively. The cyan "●" symbols represent actual reference gaze points on the monitor screen. The magenta "+" symbols represent gaze points estimated through the proposed method without considering error compensation. The blue "*" symbols represent gaze points estimated through the proposed method considering error compensation.

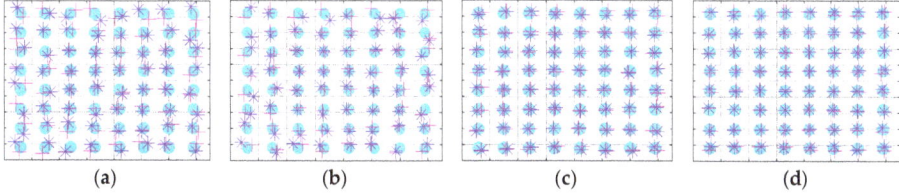

(a) (b) (c) (d)

Figure 8. Gaze point estimation with/without considering error compensation: (a) gaze point estimation utilizing a four-marker calibration model; (b) gaze point estimation utilizing a six-marker calibration model; (c) gaze point estimation utilizing a nine-marker calibration model; (d) gaze point estimation utilizing a 16-marker calibration model.

3.5. Gaze Estimation Accuracy Comparison of Different Methods

As shown in Figure 9, gaze estimation accuracy is expressed as intersection angle θ between actual gaze direction (A as gaze point) and estimated gaze direction (A' as gaze point).

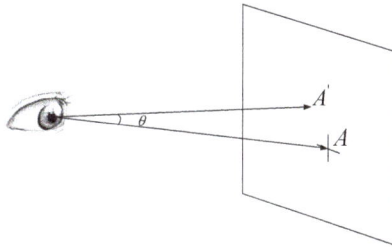

Figure 9. Definition of gaze estimation accuracy.

Angle θ can be calculated through Equation (12), where L is the distance between the human eye and the monitor screen:

$$\theta = \arctan\left(\frac{\sqrt{(x_A - x_{A'})^2 + (y_A - y_{A'})^2}}{L}\right). \tag{12}$$

The standard deviation of gaze estimation accuracy θ is defined as Equation (13), where $\bar{\theta}$ represents the mean value of θ_j ($j = 1, 2, \cdots, K$) and K is the total number of gaze points estimated:

$$\Delta_{std} = \sqrt{\frac{1}{K} \sum_{j=1}^{K} \left(\theta_j - \bar{\theta}\right)^2}. \tag{13}$$

3.5.1. Gaze Estimation Accuracy without Considering Error Compensation

Figure 10 shows a comparison of gaze estimation accuracy and standard deviation calculated through the proposed method and other neural network methods, respectively, without considering error compensation. The proposed method can provide an accuracy of 1.29°, 0.89°, 0.52°, and 0.39° when a four-, six-, nine-, or 16-marker calibration model is utilized for calibration, respectively. The maximum gaze estimation error through the proposed method for a four-, six-, nine-, or 16-marker calibration model is, respectively, 2.45°, 1.98°, 1.21°, and 0.82°. The specific results are shown in Table A1.

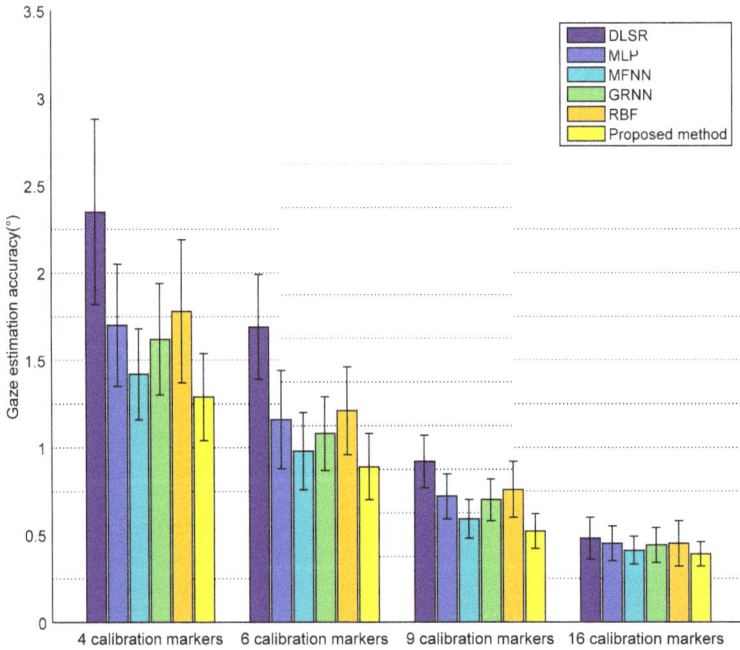

Figure 10. Comparison of gaze estimation accuracy results between proposed method and other methods without considering error compensation.

3.5.2. Gaze Estimation Accuracy Considering Error Compensation

Figure 11 shows the comparison of gaze estimation accuracy and standard deviation calculated respectively through the proposed method and other NN (Neural Network) methods considering error compensation. The proposed method can provide an accuracy of 1.17°, 0.79°, 0.47°, and 0.36° respectively, when a four-, six-, nine-, or 16-marker calibration model is utilized for calibration. When considering error compensation, the improvement percentage of gaze estimation accuracy for four-, six-, nine-, and 16-marker calibration models is 9.3%, 11.2%, 9.6%, and 7.6%, respectively. The specific results are shown in Table A2 of the Appendix.

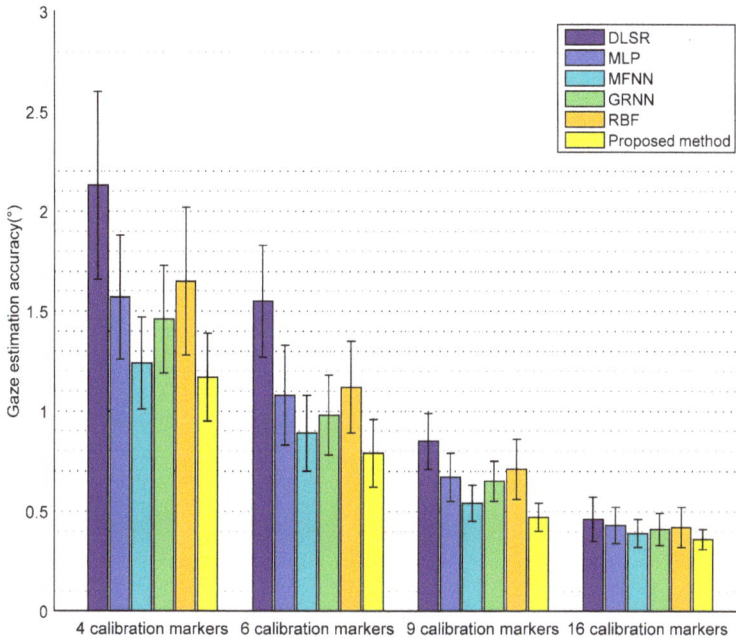

Figure 11. Comparison of gaze estimation accuracy results between proposed method and other methods considering error compensation.

4. Conclusions

In this paper, a novel 2D gaze estimation method based on pupil-glint vector is proposed on the basis of conventional gaze tracking methods. In order to realize the accurate estimation of gaze direction, an improved artificial neural network (DLSR-ANN) based on direct least squares regression is developed. Learning rate parameter, weighting function, and corresponding coefficients are determined according to trial and experience. Detected coordinates of pupil-glint vectors are applied as inputs to train an improved neural network. The mapping function model is solved and then utilized to calculate gaze point coordinates. An array of four NIR LEDs is employed to form quadrangular glints. The NIR LEDs array can generate well-distributed illumination around the human eye, which contributes to extracting pupil and glint characteristics more stably and precisely. In addition, the center coordinates of quadrangular NIR LEDs, considered as additional inputs of neural network, can further compensate for the error caused during the process of calculating the gaze point, which can enhance the accuracy of gaze point coordinates. When the gaze tracking system is established, calibration models with different numbers of markers are utilized to validate the proposed method. When a four-, six-, nine-, or 16-marker calibration model is employed for the calibration process, the proposed method can achieve an accuracy of $1.29°$, $0.89°$, $0.52°$, and $0.39°$, respectively. Taking into account error compensation, the proposed method can achieve an accuracy of $1.17°$, $0.79°$, $0.47°$, and $0.36°$, respectively, when a four-, six-, nine-, or 16-marker calibration model is employed. When considering error compensation, the improvement percentage of gaze estimation accuracy for a four-, six-, nine-, or 16-marker calibration model is 9.3%, 11.2%, 9.6%, and 7.6%, respectively. The experimental results show that the training process of the proposed method is stable. The gaze estimation accuracy of the proposed method in this paper is better than that of conventional linear regression (direct least squares regression) and nonlinear regression (generic artificial neural network). The proposed method contributes to enhance the total accuracy of a gaze tracking system.

Acknowledgments: This work is supported by Program for Changjiang Scholars and Innovation Research Team in University under Grant No. IRT1208, Basic Research Fund of Beijing Institute of Technology under Grant (No. 20130242015) and Autonomous Program of State Key Laboratory of Explosion Science and Technology under Grant No. YBKT15-09. The authors would like to thank the editor and all anonymous reviewers for their constructive suggestions.

Author Contributions: All authors made significant contributions to this article. Jianzhong Wang was mainly responsible for deployment of the system and revision of the paper; Guangyue Zhang was responsible for developing gaze estimation method, performing the experiments, and writing the paper; Jiadong Shi, the corresponding author, was responsible for conceiving and designing the experiments.

Conflicts of Interest: The authors declare no conflict of interest.

Appendix

Table A1. Comparison of gaze estimation accuracy between proposed method and other NN methods without considering error compensation.

Calibration Markers	Method	Subject 1	Subject 2	Subject 3	Subject 4	Subject 5	Subject 6	Average Err.
4	DLSR [55]	2.32° ± 0.54°	2.41° ± 0.58°	2.48° ± 0.62°	2.24° ± 0.45°	2.29° ± 0.48°	2.37° ± 0.51°	2.35° ± 0.53°
	MLP [66]	1.71° ± 0.35°	1.74° ± 0.39°	1.64° ± 0.32°	1.72° ± 0.33°	1.76° ± 0.40°	1.62° ± 0.29°	1.70° ± 0.35°
	MFNN [67]	1.38° ± 0.27°	1.45° ± 0.30°	1.43° ± 0.24°	1.49° ± 0.29°	1.34° ± 0.22°	1.40° ± 0.25°	1.42° ± 0.26°
	GRNN [68]	1.63° ± 0.32°	1.52° ± 0.28°	1.69° ± 0.35°	1.72° ± 0.36°	1.55° ± 0.30°	1.61° ± 0.33°	1.62° ± 0.32°
	RBF [69]	1.81° ± 0.43°	1.92° ± 0.48°	1.85° ± 0.44°	1.74° ± 0.37°	1.67° ± 0.33°	1.72° ± 0.41°	1.78° ± 0.41°
	Proposed	1.36° ± 0.24°	1.28° ± 0.19°	1.26° ± 0.31°	1.31° ± 0.25°	1.21° ± 0.30°	1.32° ± 0.20°	1.29° ± 0.25°
6	DLSR [55]	1.68° ± 0.29°	1.62° ± 0.25°	1.64° ± 0.28°	1.72° ± 0.31°	1.71° ± 0.33°	1.74° ± 0.31°	1.69° ± 0.30°
	MLP [66]	1.15° ± 0.26°	1.23° ± 0.33°	1.17° ± 0.25°	1.06° ± 0.24°	1.10° ± 0.27°	1.26° ± 0.35°	1.16° ± 0.28°
	MFNN [67]	0.98° ± 0.22°	0.96° ± 0.20°	1.05° ± 0.27°	1.03° ± 0.25°	0.95° ± 0.18°	0.93° ± 0.19°	0.98° ± 0.22°
	GRNN [68]	1.07° ± 0.19°	1.16° ± 0.27°	1.02° ± 0.15°	1.05° ± 0.19°	1.12° ± 0.26°	1.08° ± 0.18°	1.08° ± 0.21°
	RBF [69]	1.20° ± 0.26°	1.17° ± 0.24°	1.23° ± 0.27°	1.24° ± 0.29°	1.15° ± 0.19°	1.18° ± 0.25°	1.21° ± 0.25°
	Proposed	0.88° ± 0.16°	0.94° ± 0.19°	0.78° ± 0.25°	0.86° ± 0.14°	0.92° ± 0.21°	0.95° ± 0.18°	0.89° ± 0.19°
9	DLSR [55]	0.91° ± 0.15°	0.89° ± 0.16°	0.97° ± 0.18°	0.96° ± 0.15°	0.86° ± 0.13°	0.94° ± 0.14°	0.92° ± 0.15°
	MLP [66]	0.73° ± 0.13°	0.78° ± 0.16°	0.74° ± 0.16°	0.67° ± 0.11°	0.64° ± 0.10°	0.75° ± 0.14°	0.72° ± 0.13°
	MFNN [67]	0.58° ± 0.09°	0.57° ± 0.12°	0.64° ± 0.11°	0.56° ± 0.14°	0.59° ± 0.09°	0.62° ± 0.13°	0.59° ± 0.11°
	GRNN [68]	0.71° ± 0.11°	0.74° ± 0.12°	0.77° ± 0.16°	0.65° ± 0.09°	0.64° ± 0.10°	0.67° ± 0.12°	0.70° ± 0.12°
	RBF [69]	0.77° ± 0.17°	0.72° ± 0.14°	0.84° ± 0.21°	0.80° ± 0.20°	0.76° ± 0.15°	0.70° ± 0.12°	0.76° ± 0.16°
	Proposed	0.51° ± 0.08°	0.49° ± 0.09°	0.48° ± 0.12°	0.56° ± 0.10°	0.51° ± 0.11°	0.47° ± 0.07°	0.52° ± 0.10°
16	DLSR [55]	0.50° ± 0.12°	0.47° ± 0.10°	0.49° ± 0.13°	0.48° ± 0.15°	0.49° ± 0.09°	0.51° ± 0.14°	0.48° ± 0.12°
	MLP [66]	0.44° ± 0.11°	0.48° ± 0.13°	0.49° ± 0.11°	0.46° ± 0.09°	0.44° ± 0.10°	0.46° ± 0.08°	0.45° ± 0.10°
	MFNN [67]	0.39° ± 0.09°	0.42° ± 0.08°	0.44° ± 0.12°	0.39° ± 0.07°	0.40° ± 0.07°	0.42° ± 0.08°	0.41° ± 0.08°
	GRNN [68]	0.46° ± 0.12°	0.41° ± 0.09°	0.45° ± 0.10°	0.47° ± 0.13°	0.40° ± 0.08°	0.43° ± 0.11°	0.44° ± 0.10°
	RBF [69]	0.48° ± 0.15°	0.46° ± 0.13°	0.41° ± 0.11°	0.42° ± 0.12°	0.46° ± 0.14°	0.44° ± 0.15°	0.45° ± 0.13°
	Proposed	0.36° ± 0.06°	0.42° ± 0.09°	0.38° ± 0.08°	0.40° ± 0.07°	0.43° ± 0.08°	0.37° ± 0.06°	0.39° ± 0.07°

Table A2. Comparison of gaze estimation accuracy between proposed method and other NN methods considering error compensation.

Calibration Markers	Method	Subject 1	Subject 2	Subject 3	Subject 4	Subject 5	Subject 6	Average Err.
4	DLSR [55]	2.11° ± 0.48°	2.20° ± 0.52°	2.24° ± 0.55°	2.03° ± 0.40°	2.06° ± 0.42°	2.15° ± 0.44°	2.13° ± 0.47°
	MLP [66]	1.56° ± 0.31°	1.64° ± 0.35°	1.52° ± 0.28°	1.58° ± 0.30°	1.63° ± 0.36°	1.51° ± 0.26°	1.57° ± 0.31°
	MFNN [67]	1.23° ± 0.24°	1.21° ± 0.26°	1.28° ± 0.21°	1.26° ± 0.25°	1.18° ± 0.20°	1.25° ± 0.22°	1.24° ± 0.23°
	GRNN [68]	1.48° ± 0.29°	1.37° ± 0.22°	1.45° ± 0.31°	1.57° ± 0.28°	1.41° ± 0.26°	1.49° ± 0.28°	1.46° ± 0.27°
	RBF [69]	1.65° ± 0.39°	1.77° ± 0.43°	1.79° ± 0.40°	1.54° ± 0.34°	1.52° ± 0.29°	1.61° ± 0.36°	1.65° ± 0.37°
	Proposed	1.23° ± 0.21°	1.17° ± 0.18°	1.14° ± 0.26°	1.18° ± 0.22°	1.09° ± 0.28°	1.21° ± 0.19°	1.17° ± 0.22°
6	DLSR [55]	1.54° ± 0.28°	1.49° ± 0.23°	1.51° ± 0.26°	1.57° ± 0.30°	1.56° ± 0.31°	1.61° ± 0.29°	1.55° ± 0.28°
	MLP [66]	1.06° ± 0.24°	1.15° ± 0.30°	1.08° ± 0.21°	1.01° ± 0.22°	1.03° ± 0.23°	1.14° ± 0.29°	1.08° ± 0.25°
	MFNN [67]	0.87° ± 0.21°	0.88° ± 0.18°	0.96° ± 0.23°	0.94° ± 0.21°	0.86° ± 0.16°	0.84° ± 0.17°	0.89° ± 0.19°
	GRNN [68]	0.95° ± 0.19°	1.05° ± 0.24°	0.91° ± 0.15°	0.94° ± 0.19°	1.01° ± 0.23°	0.99° ± 0.18°	0.98° ± 0.20°
	RBF [69]	1.11° ± 0.23°	1.09° ± 0.21°	1.15° ± 0.25°	1.14° ± 0.27°	1.07° ± 0.18°	1.18° ± 0.22°	1.12° ± 0.23°
	Proposed	0.78° ± 0.13°	0.82° ± 0.17°	0.71° ± 0.23°	0.73° ± 0.12°	0.81° ± 0.20°	0.87° ± 0.17°	0.79° ± 0.17°
9	DLSR [55]	0.84° ± 0.14°	0.81° ± 0.15°	0.89° ± 0.17°	0.88° ± 0.13°	0.80° ± 0.12°	0.86° ± 0.13°	0.85° ± 0.14°
	MLP [66]	0.68° ± 0.12°	0.74° ± 0.14°	0.70° ± 0.15°	0.62° ± 0.10°	0.61° ± 0.09°	0.69° ± 0.12°	0.67° ± 0.12°
	MFNN [67]	0.53° ± 0.08°	0.52° ± 0.10°	0.60° ± 0.09°	0.51° ± 0.11°	0.54° ± 0.08°	0.56° ± 0.09°	0.54° ± 0.09°
	GRNN [68]	0.66° ± 0.09°	0.69° ± 0.11°	0.71° ± 0.15°	0.61° ± 0.08°	0.60° ± 0.09°	0.62° ± 0.10°	0.65° ± 0.10°
	RBF [69]	0.71° ± 0.16°	0.66° ± 0.13°	0.78° ± 0.18°	0.73° ± 0.19°	0.70° ± 0.14°	0.65° ± 0.11°	0.71° ± 0.15°
	Proposed	0.46° ± 0.07°	0.45° ± 0.06°	0.47° ± 0.10°	0.51° ± 0.08°	0.48° ± 0.09°	0.43° ± 0.05°	0.47° ± 0.07°
16	DLSR [55]	0.43° ± 0.09°	0.48° ± 0.12°	0.45° ± 0.10°	0.43° ± 0.09°	0.49° ± 0.13°	0.46° ± 0.10°	0.46° ± 0.11°
	MLP [66]	0.47° ± 0.11°	0.42° ± 0.09°	0.40° ± 0.08°	0.44° ± 0.07°	0.45° ± 0.10°	0.41° ± 0.11°	0.43° ± 0.09°
	MFNN [67]	0.36° ± 0.08°	0.41° ± 0.07°	0.38° ± 0.05°	0.40° ± 0.09°	0.41° ± 0.06°	0.38° ± 0.08°	0.39° ± 0.07°
	GRNN [68]	0.42° ± 0.11°	0.39° ± 0.07°	0.42° ± 0.10°	0.38° ± 0.06°	0.41° ± 0.06°	0.44° ± 0.10°	0.41° ± 0.08°
	RBF [69]	0.38° ± 0.09°	0.44° ± 0.12°	0.45° ± 0.13°	0.43° ± 0.09°	0.39° ± 0.08°	0.41° ± 0.11°	0.42° ± 0.10°
	Proposed	0.33° ± 0.05°	0.35° ± 0.04°	0.39° ± 0.06°	0.38° ± 0.08°	0.35° ± 0.04°	0.37° ± 0.05°	0.36° ± 0.05°

References

1. Jacob, R.J.K. What you look at is what you get: Eye movement-based interaction techniques. In Proceedings of the Sigchi Conference on Human Factors in Computing Systems, Seattle, WA, USA, 1–5 April 1990; pp. 11–18.

2. Jacob, R.J.K. The use of eye movements in human-computer interaction techniques: What you look at is what you get. *ACM Trans. Inf. Syst.* **1991**, *9*, 152–169.

3. Schütz, A.C.; Braun, D.I.; Gegenfurtner, K.R. Eye movements and perception: A selective review. *J. Vis.* **2011**, *11*, 89–91.

4. Miriam, S.; Anna, M. Do we track what we see? Common *versus* independent processing for motion perception and smooth pursuit eye movements: A review. *Vis. Res.* **2011**, *51*, 836–852.

5. Blondon, K.; Wipfli, R.; Lovis, C. Use of eye-tracking technology in clinical reasoning: A systematic review. *Stud. Health Technol. Inform.* **2015**, *210*, 90–94.

6. Higgins, E.; Leinenger, M.; Rayner, K. Eye movements when viewing advertisements. *Front. Psychol.* **2014**, *5*, 1–15.

7. Spakov, O.; Majaranta, P.; Spakov, O. Scrollable keyboards for casual eye typing. *Psychol. J.* **2009**, *7*, 159–173.

8. Noureddin, B.; Lawrence, P.D.; Man, C.F. A non-contact device for tracking gaze in a human computer interface. *Comput. Vis. Image Underst.* **2005**, *98*, 52–82.

9. Biswas, P.; Langdon, P. Multimodal intelligent eye-gaze tracking system. *Int. J. Hum. Comput. Interact.* **2015**, *31*, 277–294.

10. Lim, C.J.; Kim, D. Development of gaze tracking interface for controlling 3d contents. *Sens. Actuator A-Phys.* **2012**, *185*, 151–159.

11. Ince, I.F.; Jin, W.K. A 2D eye gaze estimation system with low-resolution webcam images. *EURASIP J. Adv. Signal Process.* **2011**, *1*, 589–597.

12. Kumar, N.; Kohlbecher, S.; Schneider, E. A novel approach to video-based pupil tracking. In Proceedings of the IEEE International Conference on Systems, Man and Cybernetics, Hyatt Regency Riverwalk, San Antonio, TX, USA, 11–14 October 2009; pp. 1255–1262.

13. Lee, E.C.; Min, W.P. A new eye tracking method as a smartphone interface. *KSII Trans. Internet Inf. Syst.* **2013**, *7*, 834–848.

14. Kim, J. Webcam-based 2D eye gaze estimation system by means of binary deformable eyeball templates. *J. Inform. Commun. Converg. Eng.* **2010**, *8*, 575–580.

15. Dong, L. Investigation of Calibration Techniques in video based eye tracking system. In Proceedings of the 11th international conference on Computers Helping People with Special Needs, Linz, Austria, 9–11 July 2008; pp. 1208–1215.

16. Fard, P.J.M.; Moradi, M.H.; Parvaneh, S. Eye tracking using a novel approach. In Proceedings of the World Congress on Medical Physics and Biomedical Engineering 2006, Seoul, Korea, 27 August–1 September 2006; pp. 2407–2410.

17. Yamazoe, H.; Utsumi, A.; Yonezawa, T.; Abe, S. Remote gaze estimation with a single camera based on facial-feature tracking without special calibration actions. In Proceedings of the 2008 symposium on Eye tracking research and applications, Savannah, GA, USA, 26–28 March 2008; pp. 245–250.

18. Lee, E.C.; Kang, R.P.; Min, C.W.; Park, J. Robust gaze tracking method for stereoscopic virtual reality systems. *Hum.-Comput. Interact.* **2007**, *4552*, 700–709.

19. Lee, H.C.; Lee, W.O.; Cho, C.W.; Gwon, S.Y.; Park, K.R.; Lee, H.; Cha, J. Remote gaze tracking system on a large display. *Sensors* **2013**, *13*, 13439–13463.

20. Ohno, T.; Mukawa, N.; Yoshikawa, A. Free gaze: A gaze tracking system for everyday gaze interaction. In Proceedings of the Symposium on Eye Tracking Research and Applications Symposium, New Orleans, LA, USA, 25–27 March 2002; pp. 125–132.

21. Chen, J.; Ji, Q. 3D gaze estimation with a single camera without IR illumination. In Proceedings of the International Conference on Pattern Recognition, Tampa, Florida, FL, USA, 8–11 December 2008; pp. 1–4.

22. Sheng-Wen, S.; Jin, L. A novel approach to 3D gaze tracking using stereo cameras. *IEEE Trans. Syst. Man Cybern. Part B-Cybern.* **2004**, *34*, 234–245.

23. Ki, J.; Kwon, Y.M.; Sohn, K. 3D gaze tracking and analysis for attentive Human Computer Interaction. In Proceedings of the Frontiers in the Convergence of Bioscience and Information Technologies, Jeju, Korea, 11–13 October 2007; pp. 617–621.

24. Ji, W.L.; Cho, C.W.; Shin, K.Y.; Lee, E.C.; Park, K.R. 3D gaze tracking method using purkinje images on eye optical model and pupil. *Opt. Lasers Eng.* **2012**, *50*, 736–751.

25. Lee, E.C.; Kang, R.P. A robust eye gaze tracking method based on a virtual eyeball model. *Mach. Vis. Appl.* **2009**, *20*, 319–337.

26. Wang, J.G.; Sung, E.; Venkateswarlu, R. Estimating the eye gaze from one eye. *Comput. Vis. Image Underst.* **2005**, *98*, 83–103.

27. Ryoung, P.K. A real-time gaze position estimation method based on a 3-d eye model. *IEEE Trans. Syst. Man Cybern. Part B-Cybern.* **2007**, *37*, 199–212.

28. Topal, C.; Dogan, A.; Gerek, O.N. A wearable head-mounted sensor-based apparatus for eye tracking applications. In Proceedings of the IEEE Conference on Virtual Environments, Human-computer Interfaces and Measurement Systems, Istanbul, Turkey, 14–16 July 2008; pp. 136–139.

29. Ville, R.; Toni, V.; Outi, T.; Niemenlehto, P.-H.; Verho, J.; Surakka, V.; Juhola, M.; Lekkahla, J. A wearable, wireless gaze tracker with integrated selection command source for human-computer interaction. *IEEE Trans. Inf. Technol. Biomed.* **2011**, *15*, 795–801.

30. Noris, B.; Keller, J.B.; Billard, A. A wearable gaze tracking system for children in unconstrained environments. *Comput. Vis. Image Underst.* **2011**, *115*, 476–486.

31. Stengel, M.; Grogorick, S.; Eisemann, M.; Eisemann, E.; Magnor, M. An affordable solution for binocular eye tracking and calibration in head-mounted displays. In Proceedings of the ACM Multimedia conference for 2015, Brisbane, Queensland, Australia, 26–30 October 2015; pp. 15–24.

32. Takemura, K.; Takahashi, K.; Takamatsu, J.; Ogasawara, T. Estimating 3-D point-of-regard in a real environment using a head-mounted eye-tracking system. *IEEE Trans. Hum.-Mach. Syst.* **2014**, *44*, 531–536.

33. Schneider, E.; Dera, T.; Bard, K.; Bardins, S.; Boening, G.; Brand, T. Eye movement driven head-mounted camera: It looks where the eyes look. In Proceedings of the IEEE International Conference on Systems, Man and Cybernetics, Waikoloa, HI, USA, 12 October 2005; pp. 2437–2442.

34. Min, Y.K.; Yang, S.; Kim, D. Head-mounted binocular gaze detection for selective visual recognition systems. *Sens. Actuator A-Phys.* **2012**, *187*, 29–36.

35. Fuhl, W.; Kübler, T.; Sippel, K.; Rosenstiel, W.; Kasneci, E. Excuse: Robust pupil detection in real-world scenarios. In Proceedings of the 16th International Conference on Computer Analysis of Images and Patterns (CAIP), Valletta, Malta, 2–4 September 2015; pp. 39–51.

36. Fuhl, W.; Santini, T.; Kasneci, G.; Kasneci, E. PupilNet: Convolutional Neural Networks for Robust Pupil Detection, arXiv preprint arXiv: 1601.04902. Available online: http://arxiv.org/abs/1601.04902 (accessed on 19 January 2016).

37. Fuhl, W.; Santini, T.; Kübler, T.; Kasneci, E. ElSe: Ellipse selection for robust pupil detection in real-world environments. In Proceedings of the Ninth Biennial ACM Symposium on Eye Tracking Research & Applications (ETRA' 16), Charleston, SC, USA, 14–17 March 2016; pp. 123–130.

38. Dong, H.Y.; Chung, M.J. A novel non-intrusive eye gaze estimation using cross-ratio under large head motion. *Comput. Vis. Image Underst.* **2005**, *98*, 25–51.

39. Mohammadi, M.R.; Raie, A. Selection of unique gaze direction based on pupil position. *IET Comput. Vis.* **2013**, *7*, 238–245.

40. Arantxa, V.; Rafael, C. A novel gaze estimation system with one calibration point. *IEEE Trans. Syst. Man Cybern. Part B-Cybern.* **2008**, *38*, 1123–1138.

41. Coutinho, F.L.; Morimoto, C.H. Free head motion eye gaze tracking using a single camera and multiple light sources. In Proceedings of the SIBGRAPI Conference on Graphics, Patterns and Images, Manaus, Amazon, Brazil, 8–11 October 2006; pp. 171–178.

42. Beymer, D.; Flickner, M. Eye gaze tracking using an active stereo head. In Proceedings of the IEEE Computer Society Conference on Computer Vision and Pattern Recognition, Madison, WI, USA, 18–20 June 2003; pp. 451–458.

43. Zhu, Z.; Ji, Q. Eye gaze tracking under natural head movements. In Proceedings of the IEEE Computer Society Conference on Computer Vision and Pattern Recognition, Washington, DC, USA, 20–25 June 2005; pp. 918–923.

44. Magee, J.J.; Betke, M.; Gips, J.; Scott, M.R. A human–computer interface using symmetry between eyes to detect gaze direction. *IEEE Trans. Syst. Man Cybern. Part A-Syst. Hum.* **2008**, *38*, 1248–1261.

45. Scott, D.; Findlay, J.M. *Visual Search, Eye Movements and Display Units*; IBM UK Hursley Human Factors Laboratory: Winchester, UK, 1991.

46. Wen, Z.; Zhang, T.N.; Chang, S.J. Eye gaze estimation from the elliptical features of one iris. *Opt. Eng.* **2011**, *50*.

47. Wang, J.G.; Sung, E. Gaze determination via images of irises. *Image Vis. Comput.* **2001**, *19*, 891–911.

48. Ebisawa, Y. Unconstrained pupil detection technique using two light sources and the image difference method. *WIT Trans. Inf. Commun. Technol.* **1995**, *15*, 79–89.

49. Morimoto, C.; Koons, D.; Amir, A.; Flicker, M. Framerate pupil detector and gaze tracker. In Proceedings of the International Conference on Computer Vision, Kerkyra, Greece, 20–27 September 1999; pp. 1–6.

50. Wang, C.W.; Gao, H.L. Differences in the infrared bright pupil response of human eyes. In Proceedings of Etra Eye Tracking Research and Applications Symposium, New Orleans, LA, USA, 25–27 March 2002; pp. 133–138.

51. Dodge, R.; Cline, T.S. The angle velocity of eye movements. *Psychol. Rev.* **1901**, *8*, 145–157.

52. Tomono, A.; Iida, M.; Kobayashi, Y. A TV camera system which extracts feature points for non-contact eye movement detection. In Proceedings of the SPIE Optics, Illumination, and Image Sensing for Machine Vision, Philadelphia, PA, USA, 1 November 1989; pp. 2–12.

53. Ebisawa, Y. Improved video-based eye-gaze detection method. *IEEE Trans. Instrum. Meas.* **1998**, *47*, 948–955.

54. Hu, B.; Qiu, M.H. A new method for human-computer interaction by using eye gaze. In Proceedings of the IEEE International Conference on Systems, Man, & Cybernetics, Humans, Information & Technology, San Antonio, TX, USA, 2–5 October 1994; pp. 2723–2728.

55. Hutchinson, T.E.; White, K.P.; Martin, W.N.; Reichert, K.C.; Frey, L.A. Human-computer interaction using eye-gaze input. *IEEE Trans. Syst. Man Cybern.* **1989**, *19*, 1527–1534.

56. Cornsweet, T.N.; Crane, H.D. Accurate two-dimensional eye tracker using first and fourth purkinje images. *J. Opt. Soc. Am.* **1973**, *63*, 921–928.

57. Glenstrup, A.J.; Nielsen, T.E. Eye Controlled Media: Present and Future State. Master's Thesis, University of Copenhagen, Copenhagen, Denmark, 1 June 1995.

58. Mimica, M.R.M.; Morimoto, C.H. A computer vision framework for eye gaze tracking. In Proceedings of XVI Brazilian Symposium on Computer Graphics and Image Processing, Sao Carlos, Brazil, 12–15 October 2003; pp. 406–412.

59. Morimoto, C.H.; Mimica, M.R.M. Eye gaze tracking techniques for interactive applications. *Comput. Vis. Image Underst.* **2005**, *98*, 4–24.

60. Jian-Nan, C.; Peng-Yi, Z.; Si-Yi, Z.; Chuang, Z.; Ying, H. Key Techniques of Eye Gaze Tracking Based on Pupil Corneal Reflection. In Proceedings of the WRI Global Congress on Intelligent Systems, Xiamen, China, 19–21 May 2009; pp. 133–138.

61. Feng, L.; Sugano, Y.; Okabe, T.; Sato, Y. Adaptive linear regression for appearance-based gaze estimation. *IEEE Trans. Pattern Anal. Mach. Intell.* **2014**, *36*, 2033–2046.

62. Cherif, Z.R.; Nait-Ali, A.; Motsch, J.F.; Krebs, M.O. An adaptive calibration of an infrared light device used for gaze tracking. In Proceedings of the IEEE Instrumentation and Measurement Technology Conference, Anchorage, AK, USA, 21–23 May 2002; pp. 1029–1033.

63. Cerrolaza, J.J.; Villanueva, A.; Cabeza, R. Taxonomic study of polynomial regressions applied to the calibration of video-oculographic systems. In Proceedings of the 2008 symposium on Eye Tracking Research and Applications, Savannah, GA, USA, 26–28 March 2008; pp. 259–266.

64. Cerrolaza, J.J.; Villanueva, A.; Cabeza, R. Study of polynomial mapping functions in video-oculography eye trackers. *ACM Trans. Comput.-Hum. Interact.* **2012**, *19*, 602–615.

65. Baluja, S.; Pomerleau, D. Non-intrusive gaze tracking using artificial neural networks. *Neural Inf. Process. Syst.* **1994**, *6*, 753–760.

66. Piratla, N.M.; Jayasumana, A.P. A neural network based real-time gaze tracker. *J. Netw. Comput. Appl.* **2002**, *25*, 179–196.

67. Demjen, E.; Abosi, V.; Tomori, Z. Eye tracking using artificial neural networks for human computer interaction. *Physiol. Res.* **2011**, *60*, 841–844.

68. Coughlin, M.J.; Cutmore, T.R.H.; Hine, T.J. Automated eye tracking system calibration using artificial neural networks. *Comput. Meth. Programs Biomed.* **2004**, *76*, 207–220.

69. Sesin, A.; Adjouadi, M.; Cabrerizo, M.; Ayala, M.; Barreto, A. Adaptive eye-gaze tracking using neural-network-based user profiles to assist people with motor disability. *J. Rehabil. Res. Dev.* **2008**, *45*, 801–817.

70. Gneo, M.; Schmid, M.; Conforto, S.; D'Alessio, T. A free geometry model-independent neural eye-gaze tracking system. *J. NeuroEng. Rehabil.* **2012**, *9*, 17025–17036.

71. Zhu, Z.; Ji, Q. Eye and gaze tracking for interactive graphic display. *Mach. Vis. Appl.* **2002**, *15*, 139–148.

72. Kiat, L.C.; Ranganath, S. One-time calibration eye gaze detection system. In Proceedings of International Conference on Image Processing, Singapore, 24–27 October 2004; pp. 873–876.

73. Wu, Y.L.; Yeh, C.T.; Hung, W.C.; Tang, C.Y. Gaze direction estimation using support vector machine with active appearance model. *Multimed. Tools Appl.* **2014**, *70*, 2037–2062.

74. Fredric, M.H.; Ivica, K. *Principles of Neurocomputing for Science and Engineering*; McGraw Hill: New York, NY, USA, 2001.

75. Oyster, C.W. *The Human Eye: Structure and Function*; Sinauer Associates: Sunderland, MA, USA, 1999.

76. Sliney, D.; Aron-Rosa, D.; DeLori, F.; Fankhauser, F.; Landry, R.; Mainster, M.; Marshall, J.; Rassow, B.; Stuck, B.; Trokel, S.; *et al.* Adjustment of guidelines for exposure of the eye to optical radiation from ocular instruments: Statement from a task group of the International Commission on Non-Ionizing Radiation Protection. *Appl. Opt.* **2005**, *44*, 2162–2176.

77. Wang, J.Z.; Zhang, G.Y.; Shi, J.D. Pupil and glint detection using wearable camera sensor and near-infrared led array. *Sensors* **2015**, *15*, 30126–30141.

78. Lee, J.W.; Heo, H.; Park, K.R. A novel gaze tracking method based on the generation of virtual calibration points. *Sensors* **2013**, *13*, 10802–10822.

79. Gwon, S.Y.; Jung, D.; Pan, W.; Park, K.R. Estimation of Gaze Detection Accuracy Using the Calibration Information-Based Fuzzy System. *Sensors* **2016**.

A Modified Feature Selection and Artificial Neural Network-Based Day-Ahead Load Forecasting Model for a Smart Grid

Ashfaq Ahmad, Nadeem Javaid, Nabil Alrajeh, Zahoor Ali Khan, Umar Qasim and Abid Khan

Abstract: In the operation of a smart grid (SG), day-ahead load forecasting (DLF) is an important task. The SG can enhance the management of its conventional and renewable resources with a more accurate DLF model. However, DLF model development is highly challenging due to the non-linear characteristics of load time series in SGs. In the literature, DLF models do exist; however, these models trade off between execution time and forecast accuracy. The newly-proposed DLF model will be able to accurately predict the load of the next day with a fair enough execution time. Our proposed model consists of three modules; the data preparation module, feature selection and the forecast module. The first module makes the historical load curve compatible with the feature selection module. The second module removes redundant and irrelevant features from the input data. The third module, which consists of an artificial neural network (ANN), predicts future load on the basis of selected features. Moreover, the forecast module uses a sigmoid function for activation and a multi-variate auto-regressive model for weight updating during the training process. Simulations are conducted in MATLAB to validate the performance of our newly-proposed DLF model in terms of accuracy and execution time. Results show that our proposed modified feature selection and modified ANN (m(FS + ANN))-based model for SGs is able to capture the non-linearity(ies) in the history load curve with 97.11% accuracy. Moreover, this accuracy is achieved at the cost of a fair enough execution time, *i.e.*, we have decreased the average execution time of the existing FS + ANN-based model by 38.50%.

Reprinted from *Appl. Sci.* Cite as: Ahmad, A.; Javaid, N.; Alrajeh, N.; Khan, Z.A.; Qasim, U.; Khan, A. A Modified Feature Selection and Artificial Neural Network-Based Day-Ahead Load Forecasting Model for a Smart Grid. *Appl. Sci.* **2016**, *5*, 1756–1772.

1. Introduction

On a customer service platform, the physical power system along with information and communication technology that link together heterogeneous devices in an automated fashion to improve the parameters of interest is a smart grid (SG) (refer to Figure 1 [1]). It is more likely that the SG will integrate new communication technologies, advanced metering, distributed systems, distributed storage, security and safety to achieve considerable robustness and reliability [2–4].

Two-way communication is one of the key enablers that turns a traditional power grid into a smart one, based on which optimal decisions are made by the energy management unit [2]. In this regard, many demand-side scheduling techniques are proposed [5–8]. However, there exists sufficient challenges prior to scheduling techniques in terms of stochastic information schemes to predict the future load. Thus, with the growing expectation of the adoption of SGs, advanced techniques and tools are required to optimize the overall operation.

Day-ahead load forecasting (DLF) is one of the fundamental, as well as essential tasks that is needed for proper operation of the SG. On another note, accurate load forecasting leads to enhanced management of resources (renewable and conventional), which in turn directly affects the economies of the energy trade. However, in terms of DLF, the SG is more difficult to realize due to lower similarities (high randomness due to more load fluctuations) in the history load curves as compared to that of long-term load forecasting. In the literature, many attempts have been made to develop an accurate DLF model for SGs. For example, a bi-level DLF strategy is presented in [9]; however, this strategy is very complex in terms of implementation, which leads to a high execution time. Similarly, another load forecasting model based on a Gaussian process is presented in [10], which is not complex in terms of implementation; however, this model pays the cost of accuracy to achieve relatively less execution time. The model proposed in [11] focuses on day-ahead load forecasting in energy-intensive enterprises; however, this model is very complex, and thus, its execution time is relatively on the higher side.

As mentioned earlier, the day-ahead load of an SG shows more fluctuations as compared to its long-term load. Accurate DLF model development with a fair enough execution time in these SGs is thus a highly challenging task. Alternatively, DLF accuracy enhancement may be achieved to some extent, however, at the cost of execution time. Therefore, we focus on the development of an accurate enough DLF model with a fair enough execution time for SGs. Our proposal consists of three modules: the data preparation module, the feature selection module and the forecast module. The first module normalizes and then encodes the input historical load data. This encoded information is sent to the feature selection module, where redundant and irrelevant features are removed from the input load data. It is worth mentioning here that in the feature selection module, we use our modified version of the famous mutual information technique (a detailed discussion is provided in Section 3.2). The selected features are sent to the ANN-based forecast module, which uses a sigmoid function for activation and a multi-variate auto-regressive model for weight updating during the training process. In simulations, we compare our newly-proposed model with an existing one in terms of forecast accuracy and execution time. Results justify the applicability of our proposition.

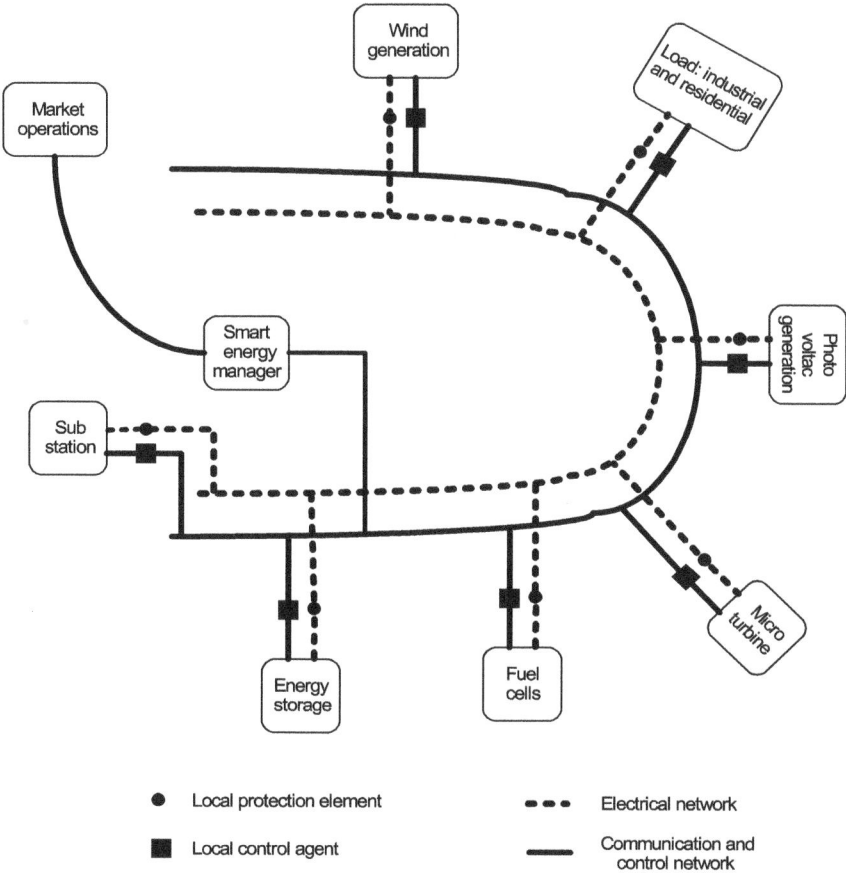

Figure 1. A smart grid (SG).

The rest of the paper is organized as follows. Section 2 contains relevant DLF contributions from the research community. Section 3 provides a brief description of the proposed ANN-based DLF model for SGs. Section 4 provides the discussion of the simulation results, and Section 5 ends the research work with conclusions and future work.

2. Related Work

Since accurate load forecasting is directly related to the economies of the energy trade, in this regard, we discuss some previous load forecasting attempts in SGs as follows.

In [9], the authors study the characteristics of the load time series of an SG and then compare its differences with that of a traditional power system. In addition, the authors propose a bi-level (upper and lower) short-term load prediction strategy for SGs. The lower level is a forecaster that utilizes a neural network and evolutionary algorithm. The upper level optimizes the performance of the lower level by using the differential evolution algorithm. In terms of effectiveness, the proposed bi-level prediction strategy is evaluated via real-time data of a Canadian university. This work is very effective in terms of accuracy; however, its execution time is very high. (Note: in the simulations, we have compared [12] with our proposed work. Results show that our proposed model takes 38.50% less time to execute than the work in [12]. The work in [9] adds an evolutionary algorithm-based module to the work in [12]. This means that [9] will take more time to execute than [12]. That is why we have stated the very high execution of [9].)

In [10], the authors develop a DLF model that is based on a Gaussian process. The proposed predictive methodology captures the heteroscedasticity of load in an efficient manner. In addition, they overcome the computational complexity of the Gaussian process by using a $\frac{1}{2}$ regularizer. A simulation-based study is carried out to prove the effectiveness of the proposed model. The authors have overcome the complexity of the Gaussian distribution to some extent; however, the future predictions are still highly questionable in terms of accuracy.

In [11], a probabilistic approach is presented to generate the energy consumption profile of household appliances. The proposed approach takes a wide range of appliances into consideration along with a high degree of flexibility. Moreover, this approach configures the households between working days and holidays by utilizing the Gaussian distribution-based methodology. However, due to the absence of a closed form solution of the Gaussian distribution, the algorithm is very complex. Moreover, the authors assume a Gaussian distribution not only for the number of active devices in a home, but also for their power usage. These assumption are not always true, thereby making future predictions highly questionable in terms of accuracy.

An artificial neural network-based short-term load forecasting method is presented in [13]. The proposed methodology is divided into four steps. Step 1 deals with the techniques of data selection. Step 2 is for wavelet transform. Step 3 is based on ANN-based forecasting. Step 4 takes into consideration the error-correcting functions. The effectiveness of the proposed methodology is verified by using practical household load demands. This algorithm has better accuracy than the aforementioned ones; however, accuracy is achieved at the cost of execution time.

A stochastic model for tackling the load fluctuations of users is presented in [14], which is robust enough to predict load. This work exploits Markov chains to capture stochasticity associated with user's energy consumption in a heterogeneous environment. In other words, the authors exploit information associated with the daily activities of users to predict their future demand. In this scheme, the future predictions do not depend on past values; that not only makes it robust, but also relatively less complex, however at the cost of accuracy.

A novel technique for price spike occurrence prediction is presented in [15]. This model is comprised of two modules; wavelet transform for feature selection and ANNs to predict the future price spikes. Irrelevant and redundant data are discarded from the input dataset, such that the selected inputs are fed into the probabilistic neural network-based forecaster. The authors evaluate their proposed method using real-time data from the PJM and Queensland electricity markets. This technique is accurate; however, wavelet transform for feature selection makes it relatively more complex.

In [12], the authors use a combination of a mutual information-based feature selection technique and a cascaded neuro-evolutionary algorithm to predict the day-ahead price of electricity markets. They also incorporate an iterative search procedure to fine-tune the adjustable parameters of both the neuro-evolutionary algorithm and the feature selection technique. The combination of various techniques makes this algorithm efficient in terms of accuracy, however at the cost of execution time.

3. Our Proposed Work

Subject to the complex day-ahead load forecast of SGs, any proposed prediction strategy should be capable enough to mitigate the non-linear input/output relationship as efficiently as possible. We choose an ANN-based forecaster for two reasons; (i) these can capture non-linearity in historical load data; and (ii) the flexibility and ease in implementation with acceptable accuracy (note: both of these reasons are justified via simulations). However, prior to ANN-based forecasting, input load time series must be made compatible. Therefore, our proposed day-ahead load forecasting model (for SGs) consists of three modules: the data preparation module, the feature selection module and the forecast module (refer to Figure 2). The first module performs pre-processing to make the input data compatible with the feature selection module and the forecast module. The second module removes irrelevant and redundant features from the input data. The third module consists of an ANN to forecast the day-ahead load of the SG. The details are as follows.

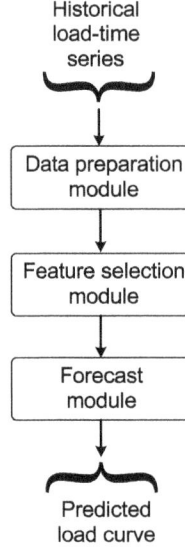

Figure 2. Block diagram of the proposed methodology.

3.1. Pre-Processing Module

Suppose that the input load time series is shown by the following matrix:

$$
P = \begin{bmatrix}
p_{h_1}^{d_1} & p_{h_2}^{d_1} & p_{h_3}^{d_1} & \cdots & p_{h_m}^{d_1} \\
p_{h_1}^{d_2} & p_{h_2}^{d_2} & p_{h_3}^{d_2} & \cdots & p_{h_m}^{d_2} \\
p_{h_1}^{d_3} & p_{h_2}^{d_3} & p_{h_3}^{d_3} & \cdots & p_{h_m}^{d_3} \\
\vdots & \vdots & \vdots & \ddots & \vdots \\
p_{h_1}^{d_n} & p_{h_2}^{d_n} & p_{h_3}^{d_n} & \cdots & p_{h_m}^{d_n}
\end{bmatrix}
\tag{1}
$$

where h_m is the m-th hour, d_n is the n-th day and $p_{h_m}^{d_n}$ is the historical power consumption value at the m-th hour of the n-th day. As there are 24 h in a day, $m = 24$. The value of n depends on the designer's choice, i.e., a greater value of n leads to fine tuning during the training process of the forecast module, because more lagged samples of input data are available. However, this would lead to greater execution time.

Prior to feeding the feature selection module with input matrix P, the following step-wise operations are performed by the data preparation module (refer to Figure 3):

1. Local maximum: Initially, a local maximum value is calculated for each column of the P matrix; $p_{max}^{c_i} = max\{p_{h_i}^{d_1}, p_{h_i}^{d_2}, p_{h_i}^{d_3}, \dots, p_{h_i}^{d_n}\}, \forall\, i \in \{1, 2, 3, \dots, n\}$.

196

2. Local normalization: In this step, each column of the matrix P is normalized by its respective local maxima, such that the resultant matrix is represented by P_{nrm}. Now, each entry of P_{nrm} ranges between zero and one.

3. Local median: For each column of the P_{nrm} matrix, a local median value Med_i is calculated ($\forall i \in \{1, 2, 3, \ldots, n\}$).

4. Binary encoding: Each entry of the P_{nrm} matrix is compared to its respective Med_i value. If the entry is less than its respective local median value, then it is encoded with a binary zero; else, it is encoded with a binary one. In this way, a resultant matrix containing only binary values (zeroes and ones), P_b, is obtained.

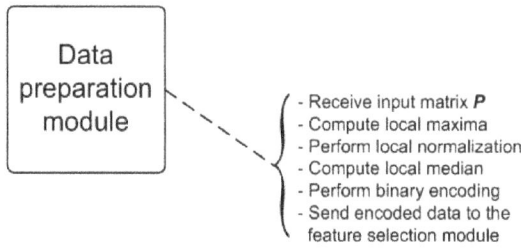

Figure 3. Data preparation module.

Note: the load/consumption pattern is different for different days, *i.e.*, the load pattern on holidays is different from that on working days. In order to enhance the accuracy of prediction strategy, the training samples must be relevant. Similarly, a lesser number of training samples will decrease the execution time of the prediction strategy. The above two reasons lead us to prefer local normalization over global normalization.

At this stage, the P_b matrix is compatible with the feature selection module and is thus fed into it.

3.2. Feature Selection Module

Once the data are binary encoded, not only redundant, but also irrelevant samples need to be removed from the lagged input data samples. In removing redundant features, the execution time during the training process is minimized. On the other hand, removal of irrelevant features leads to improvement in forecast accuracy, because the outliers are removed.

In order to remove the irrelevant and redundant features from the binary encoded input data matrix P_b, an entropy-based mutual information technique is used in [9,12], which defines the mutual information between input Q and target T by the following formula,

197

$$MI(Q,T) = \sum_i \sum_j p(Q_i, T_j) log_2 \left(\frac{p(Q_i, T_j)}{p(Q_i)p(T_i)} \right) \quad \forall i, j \in \{0,1\} \tag{2}$$

In Equation (2), $MI(Q,T) = 0$ means that Q and T are independent; a high value of $MI(Q,T)$ means that Q and T are strongly related, and a low value of $MI(Q,T)$ means that Q and T are loosely related.

Thus, the candidate inputs are ranked with respect to the mutual information value between input and target values. In [9,12], the target values are chosen as the last samples for every hour of the day among all of the training samples (for every hour, only one target value is chosen that is the value of the previous day). The choice of the last sample seems logical, as it is the closest value to the upcoming day with respect to time; however, it may lead to serious forecast errors due to the lack of consideration of the average behaviour. However, consideration of only the average behaviour is also insufficient, because the last sample has its own importance. To sum up, we come up with a solution that not only considers the last sample, but also the average behaviour. Thus, we modify Equation (2) for three discrete random variables as,

$$MI(Q,T,M) = \sum_i \sum_j \sum_k p(Q_i, T_j, M_k) log_2 \left(\frac{p(Q_i, T_j, M_k)}{p(Q_i)p(T_i)p(M_k)} \right) \quad \forall i, j \in \{0,1\} \tag{3}$$

In expanded form, Equation (3) is written as follows,

$$
\begin{aligned}
MI(Q,T,M) = {} & p(Q=0,T=0,M=0) \times log_2 \left(\frac{p(Q=0,T=0,M=0}{p(Q=0)p(T=0)p(M=0)} \right) \\
& + p(Q=0,T=0,M=1) \times log_2 \left(\frac{p(Q=0,T=0,M=1}{p(Q=0)p(T=0)p(M=1)} \right) \\
& + p(Q=0,T=1,M=0) \times log_2 \left(\frac{p(Q=0,T=1,M=0}{p(Q=0)p(T=1)p(M=0)} \right) \\
& + p(Q=0,T=1,M=1) \times log_2 \left(\frac{p(Q=0,T=1,M=1}{p(Q=0)p(T=1)p(M=1)} \right) \\
& + p(Q=1,T=0,M=0) \times log_2 \left(\frac{p(Q=1,T=0,M=0)}{p(Q=1)p(T=0)p(M=0)} \right) \\
& + p(Q=1,T=0,M=1) \times log_2 \left(\frac{p(Q=1,T=0,M=1)}{p(Q=1)p(T=0)p(M=1)} \right) \\
& + p(Q=1,T=1,M=0) \times log_2 \left(\frac{p(Q=1,T=1,M=0)}{p(Q=1)p(T=1)p(M=0)} \right) \\
& + p(Q=1,T=1,M=1) \times log_2 \left(\frac{p(Q=1,T=1,M=1)}{p(Q=1)p(T=1)p(M=1)} \right)
\end{aligned}
\tag{4}
$$

In order to determine the *MI* value between Q and T, the joint and independent probabilities need to be determined. For this purpose, an auxiliary variable A_v is introduced.

$$A_v = 4T + 2M + Q \quad \forall T, M, Q \in \{0,1\} \tag{5}$$

It is clear from Equation (5) that A_v ranges between zero and seven. $A_{0v}, A_{1v}, A_{2v}, A_{3v}, ..., A_{7v}$ counts the number of sample data points (out of total l data points) for which $A_v = 0$, $A_v = 1$, $A_v = 2$, $A_v = 3,...$, $A_v = 7$, respectively. In this way, we can now easily determine the joint and independent probabilities as follows.

$$p(Q = 0, T = 0, M = 0) = \frac{A_{0v}}{l}$$

$$p(Q = 0, T = 0, M = 1) = \frac{A_{2v}}{l}$$

$$p(Q = 0, T = 1, M = 0) = \frac{A_{4v}}{l}$$

$$p(Q = 0, T = 1, M = 1) = \frac{A_{6v}}{l} \tag{6}$$

$$p(Q = 1, T = 0, M = 0) = \frac{A_{1v}}{l}$$

$$p(Q = 1, T = 0, M = 1) = \frac{A_{3v}}{l}$$

$$p(Q = 1, T = 1, M = 0) = \frac{A_{5v}}{l}$$

$$p(Q = 1, T = 1, M = 1) = \frac{A_{7v}}{l}$$

$$p(Q = 0) = \frac{A_{0v} + A_{2v} + A_{4v} + A_{6v}}{l}$$

$$p(Q = 1) = \frac{A_{1v} + A_{3v} + A_{5v} + A_{7v}}{l}$$

$$p(T = 0) = \frac{A_{0v} + A_{1v} + A_{2v} + A_{3v}}{l}$$

$$p(T = 1) = \frac{A_{4v} + A_{4v} + A_{5v} + A_{7v}}{l} \tag{7}$$

$$p(M = 0) = \frac{A_{0v} + A_{1v} + A_{4v} + A_{5v}}{l}$$

$$p(M = 1) = \frac{A_{2v} + A_{3v} + A_{6v} + A_{7v}}{l}$$

Based on Equation (4), mutual information between Q and T is calculated, and thus, redundancy and irrelevancy are removed from the input samples. This mutual information-based technique is computed with a reasonable execution time and acceptable accuracy.

3.3. Forecast Module

By evaluating load variations over several months, or between two consecutive days, or between consecutive hours over a day, [16] concluded that SG's load time series signal exhibits strong volatility and randomness. This result is obvious, because different users have different energy/power consumption patterns/habits. Thus, in terms of DLF, realization of an SG is more difficult as compared to its realization in terms of long-term load forecast. Therefore, the basic requirement of the forecast module is to forecast the load time series of an SG by taking into consideration its non-linear characteristics. In this regard, ANNs are widely used for two reasons; accurate forecast ability and the ability to capture the non-linear characteristics.

Due to the aforementioned reasons, we choose an ANN-based implementation in our forecast module. Initially, the forecast module receives selected features $SF(.)$ and then constructs training "TS" and validation samples "VS" from it as follows:

$$
\begin{aligned}
TS &= SF(i,j), \quad \forall i \in \{2,3,\ldots,m\} \\
&\text{and} \quad \forall j \in \{1,2,3,\ldots,n\}
\end{aligned}
\tag{8}
$$

$$
VS = SF(1,j), \quad \forall j \in \{1,2,3,\ldots,n\}
\tag{9}
$$

From Equations (8) and (9), it is clear that the ANN is trained by all of the historical load time series candidates, except the last one, which is used for validation purpose. This discussion leads us towards the explanation of the training mechanism. However, prior to the explanation, it is essential to describe the ANN.

An ANN, inspired by the nervous system of humans, is a set of artificial neurons (ANs) to perform the tasks of interest (note: our task of interest is the DLF of SGs). Usually, an AN performs a non-linear mapping from $\mathbb{R}^I$ to $[0,1]$ that depends on the activation function used.

$$
f_{act}^{AN} : \mathbb{R}^I \rightarrow [0,1]
\tag{10}
$$

where I is the vector of the input signal to the AN (here, inputs are the selected features only). Figure 4 illustrates the structure of an AN that receives $I = (I_1, I_2, \ldots, I_n)$. In order to either deplete or strengthen the input signal, to each I_i is associated a weight w_i. The ANN computes I and uses f_{act}^{AN} to compute the output signal "y". However, the strength of y is also influenced by a bias value (threshold) "b". Therefore, we can compute I as follows:

$$
I = \sum_{i=1}^{i_{max}} I_i\, w_i
\tag{11}
$$

The f_{act}^{AN} receives I and b to determine y. Generally, f_{act}^{AN}'s are mappings that monotonically increase ($f_{act}^{AN}(-\infty = 0)$ and $f_{act}^{AN}(+\infty = 1)$). Among the typically used f_{act}^{AN}'s, we use sigmoid f_{act}^{AN}.

$$f_{act}^{AN}(I, b) = \frac{1}{1 + e^{-\alpha(I-b)}} \tag{12}$$

We choose sigmoid f_{act}^{AN} due for two reasons; $f_{act}^{AN} \in (0,1)$, and the parameter α has the ability to control the steepness of the f_{act}^{AN}. In other words, the sigmoid f_{act}^{AN} choice enables the AN to capture the non-linear characteristic of load time series. Since this work aims at the DLF for SGs, and one day consists of 24 h, the ANN consists of 24 forecasters (one AN for an hour), where each forecaster predicts the load of one hour of the next day. In other words, 24 hourly load time series are separately modelled instead of one complex forecaster. The whole process is repeated every day to forecast the load of the next day.

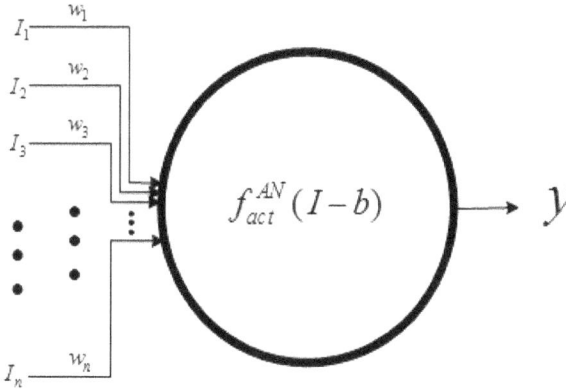

Figure 4. An artificial neuron.

The question that now needs to be answered is how to determine w_i and b? The answer is straight forward, $i.e.$, via learning. In our case, prior knowledge of load-time series exists. Thereby, we use supervised learning; adjusting w_i and b values until a certain termination criterion is satisfied. The basic objective of supervised training is to adjust w_i and b such that the error signal "$e(k)$" between the target value "$\hat{y}(k)$" and real output of neuron "$y(k)$" is minimized.

$$\begin{aligned} Minimize \quad e(k) &= y(k) - \hat{y}(k), \\ \forall k &\in \{1,2,3,\ldots,m\} \end{aligned} \tag{13}$$

We use the method of least squares to determine the parameter matrices, which is given as follows,

$$Minimize \quad J = \sum_{k=1}^{m} e^T(k)e(k),$$
$$\forall k \in \{1,2,3,\ldots,m\} \tag{14}$$

Subject to the most feasible solution of Equation (14), we use the multi-variate auto-regressive model presented in [17], because it solves the objective function in relatively less time with reasonable accuracy, as compared to the typically used learning rules, like gradient descent, Widrow–Hoff and delta [18]. According to [17], the parameter matrices are given as follows,

$$\sum_{i=1}^{n} W(i)R(j-i) = 0, \quad j = \{2,3,\ldots,n\} \tag{15}$$

$$\sum_{i=1}^{n} \overline{W}(i)R(i-j) = 0, \quad j = \{2,3,\ldots,n\} \tag{16}$$

where $W(1) = I_D$ (I_D is the identity matrix), $\overline{W}(1) = I_D$ and R is the cross co-relation given as:

$$R(i) = \frac{1}{n} \sum_{k=i}^{n-1-i} [x(k)-m][x(k-i)-m]^T \tag{17}$$

In Equation (11), m is the mean vector of the observed data,

$$m = \frac{1}{n} \sum_{k=i}^{n} x(k) \tag{18}$$

Based on these equations, [17] defines the following prediction error co-variance matrices.

$$\left.\begin{array}{l} V_t = \sum_{k=1}^{n} W_t(k)R(-k) \\ \overline{V}_t = \sum_{k=1}^{n} \overline{W}_t(k)R(-k) \\ \Delta_t = \sum_{k=1}^{n} W_t(-k)R(t-k+1) \\ \overline{\Delta}_t = \sum_{k=1}^{n} \overline{W}_t(k)R(-t+k-1) \end{array}\right\} \tag{19}$$

The recursive equations are as follows:

$$\left.\begin{array}{l} W_{t+1}(k) = W_t(k)W_{t+1}(t+1)\overline{W}_t(t-k+1) \\ \overline{W}_{t+1}(k) = \overline{W}_t(k)\overline{W}_{t+1}(t+1)\overline{W}_t(t-k+1) \end{array}\right\} \tag{20}$$

$$\left. \begin{array}{l} W_{t+1}(t+1) = -\Delta_t \overline{V}_t^{-1} \\ \overline{W}_{t+1}(t+1) = -\overline{\Delta}_t V_t^{-1} \end{array} \right\} \tag{21}$$

In order to find the weights, Equations (20) and (21) are solved recursively. For further details about the weight update mechanism, Equations (15)–(21), readers are suggested to read [17]. Figure 5 is a pictorial representation of the steps involved in the data forecast module.

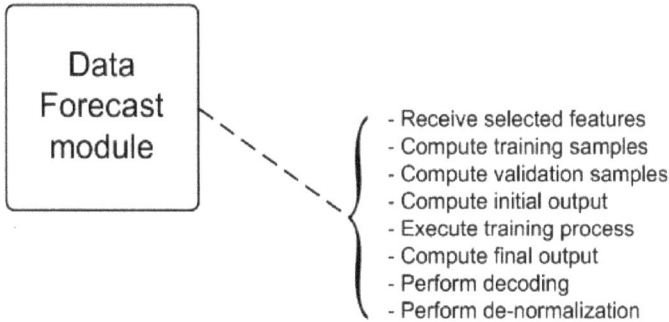

Data
Forecast
module

- Receive selected features
- Compute training samples
- Compute validation samples
- Compute initial output
- Execute training process
- Compute final output
- Perform decoding
- Perform de-normalization

Figure 5. Data forecast module.

Once the weights in Equations (20) and (21) are recursively adjusted as per the objective function in Equation (13), the output matrix is then binary decoded and de-normalized to get the desired load time series. The stepwise algorithm of the proposed methodology is shown in Algorithm 1.

Note: our proposed prediction model predicts tomorrow's load on the basis of historical load till today. Thus, the prediction model never fails, *i.e.*, for every next day, the model needs information till the current day. However, the proposed model is unable to predict the load for more than tomorrow provided the historical load information till today.

Algorithm 1 Day-ahead load forecast.

1: **Pre-conditions:** $i =$ number of days, and $j =$ number of hours per day
2: $P \leftarrow$ historical load data
3: Compute $P_{max}^{c_i} \quad \forall i \in \{1, 2, 3, \ldots, n\}$
4: Compute P_{nrm}
5: Compute $Med_i \quad \forall i \in \{1, 2, 3, \ldots, n\}$
6: **for all** $(i \in \{1, 2, 3, \ldots, n\})$ **do**
7:     **for all** $(j \in \{1, 2, 3, \ldots, m\})$ **do**
8:       **if** $(P_{nrm}^{(i,j)} \leq Med_i)$ **then**
9:         $P_b^{i,j} \leftarrow 0$
10:       **else if then**
11:         $P_b^{i,j} \leftarrow 1$
12:       **end if**
13:     **end for**
14: **end for**
15: Remove redundant and irrelevant features using Equation (4)
16: Compute TS and VS using Equations (8) and (9), respectively
17: Compute $y(1)$ by letting $W(1) = I$ and
    $\overline{W}(1) = I$
18: **while** Maximum number of iterations not reached **do**
19:     **if** $J(k+1) \leq J(k)$ **then**
20:       $y(k) \leftarrow y(k+1)$
21:     **else if then**
22:       Train ANN as per Equations (20) and (21)
23:       Compute $y(k+1)$ and go back to Step (18)
24:     **end if**
25: **end while**
26: Perform decoding
27: Perform de-normalization
---

4. Simulation Results

We evaluate our proposed DLF model (m(MI + ANN)) by comparing it with an existing MI + ANN model in [12]. We choose the existing MI + ANN model in [12] for comparison, because its architecture has a close resemblance to our proposed model. In our simulations, historical load time series data from November (2014) to January (2015) are taken from the publicly-available PJM electricity market for two SGs in the United States of America; DAYTOWN and EKPC [19]. November to December (2014) data are used for training and validation purposes, and January (2015) data are used for testing purposes. Simulation parameters are shown in Table 1, and their justification can be found in [9,12,17,18]. In this paper, we have considered two performance metrics; % error and execution time (convergence rate).

- Error performance: This is the difference between the actual and the forecast signal/curve and is measured in %.
- Convergence rate or execution time: This is the simulation time taken by the system to execute a specific forecast model. Forecast models for which the execution time is small are said to converge quickly as compared to the opposite case. In this paper, execution time is measured in seconds.

Table 1. Simulation parameters.

Parameter	Value
Number of forecasters	24
Number of hidden layers	1
Number of neurons in the hidden unit	5
Number of iterations	100
Momentum	0
Initial weights	0.1
Historical load data	26 days
Bias value	0

Figures 6a and 7a are the graphical illustrations of how well our proposed ANN-based DALF model predicts the target values of an SG. In these figures, the proposed m(MI + ANN)-based forecast curve more tightly follows the target curve as compared to the existing MI + ANN-based forecast curve, which is justification of the theoretical discussion of our proposed methodology in terms of non-linear forecast ability. Not only the sigmoid f_{act}^{AN} (refer to equation), but also the multivariate auto-regressive training algorithm enable the day-ahead ANN-based forecast methodology to capture non-linearity(ies) in historical load data.

Figure 6b shows the % forecast error when tests are conducted on the DAYTOWN grid; our m(MI + ANN) forecasts with 2.9% and the existing MI + ANN forecasts with 3.84% relative errors, respectively. Similarly, Figure 7b shows the % forecast error when tests are conducted on the EKPC grid; our m(MI + ANN) forecasts with 2.88% and the existing MI + ANN forecasts with 3.88% relative errors, respectively. This improvement in terms of relative % error performance by our proposed DALF model is due to the following two reasons: (i) the modified feature selection technique in our proposed DALF model; and (ii) multi-variate auto-regressive training algorithm. The first reason accounts for the removal of redundant, as well as irrelevant features from the input data in a more efficient way as compared to the existing DALF model. By a more efficient way, we mean that as our proposal considers the average sample in the feature selection process, as well in addition to the last sample and the target sample. Thus, the margin of outliers that cause significant relative % error is down-sized. The second reason deals with the selection of an efficient training algorithm, as our proposition trains the ANN via the multi-variate auto-regressive algorithm and the existing DALF model trains the ANN via Levenberg–Marquardt algorithm.

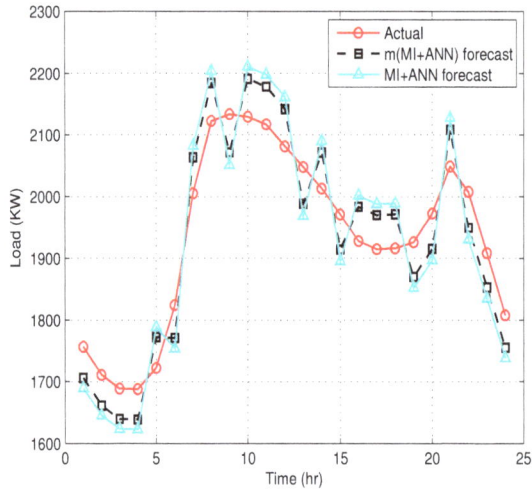

(b)

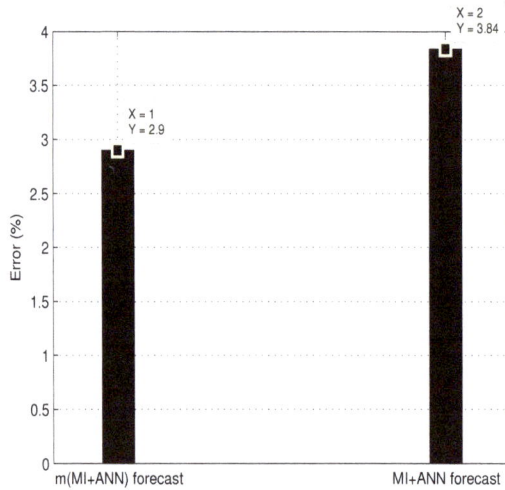

(c)

Figure 6. *Cont.*

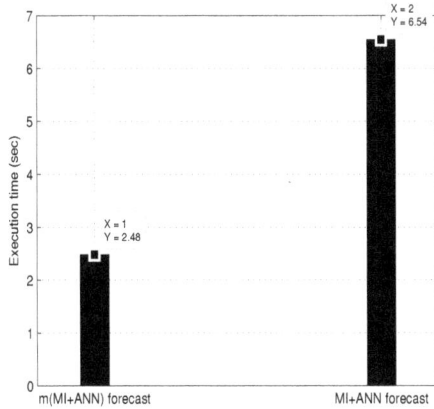

(a)

Figure 6. DAYTOWN (27 January 2015): m(MI + ANN) forecast *vs.* MI + ANN forecast. (**a**) Actual *vs.* forecast; (**b**) error performance; (**c**) convergence rate analysis.

As discussed in Sections 1, 2 and 3 that there exist a trade-off between forecast accuracy and execution time. However, Figures 6b,c and 7b,c show that our proposed DALF model not only results in relatively less % error but also less execution time. As mentioned earlier, our devised modifications in the feature selection process and selection of the multi variate training algorithm cause relative improvement in terms of % error. On the other hand, m(MI + ANN) model converges with a faster rate (less execution time) as compared to the existing MI + AN model due to three reasons; (i) exclusion of the local optimization algorithm subject to error minimization; (ii) modified feature selection process; and (iii) selection of multi variate auto regressive training algorithm. Quantitatively (Figures 6c and 7c), the execution time of existing model is 6.54 s for DAYTOWN grid and 6.60 s for EKPC grid, and that of our proposed model is 2.48 s for DAYTOWN and 2.58 s for EKPC, respectively. In these figures, the relative improvement in execution time is 37.92% for DAYTOWN, 39.09% for EKPC. Our proposition selects features from the input data while considering average sample, last sample and the target sample. This means that the chances of outliers in selected features have been significantly decreased, and the local optimization algorithm used by the existing MI + ANN forecast model is not further needed. Our proposed m(MI + ANN) forecast model does not account for the execution time taken by the iterative optimization algorithm. As a result, our proposed DALF model converges with a faster rate as compared to the existing DALF model.

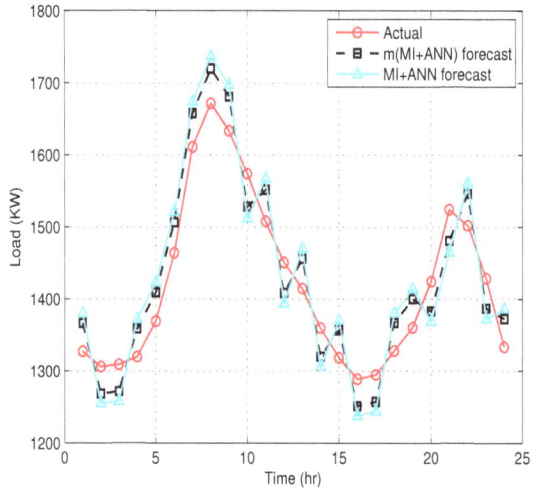

(b)

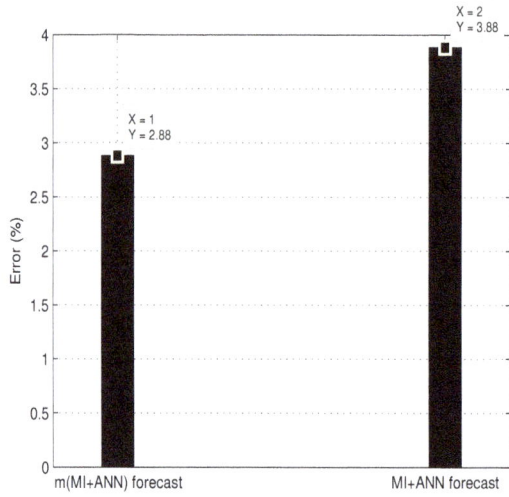

(c)

Figure 7. *Cont.*

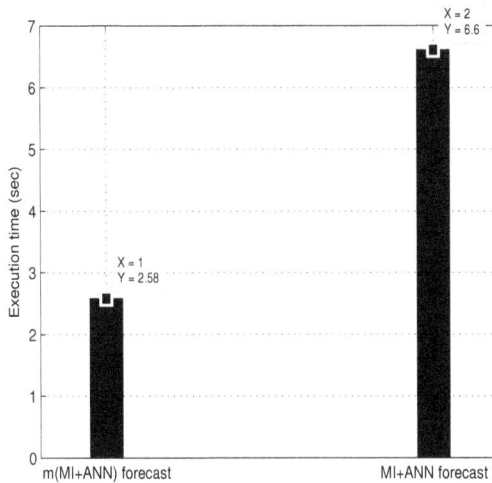

(a)

Figure 7. EKPC (27 January 2015): m(MI + ANN) forecast *vs.* MI + ANN forecast. (a) Actual *vs.* forecast; (b) Error performance; (c) Convergence rate analysis.

5. Conclusion and Future Work

In SGs, current research work primarily focuses on optimization techniques of power scheduling. However, prior to scheduling, an accurate load forecasting model is needed, because accurate load forecasting leads to enhanced management of resources, which in turn directly affects the economies of the energy trade. Furthermore, lower similarities (high randomness) and non-linearity in history load curves make the SG's DLF more challenging as compared to long-term load forecasting. Thus, the aforementioned reasons lead us to investigate the SG's DLF models. From a literature review, we found that many DLF models are proposed for SGs; however, these models trade off between accuracy and execution time. Thus, we focus on the development of an accurate DLF model with reduced execution time. In this regard, this paper has presented an ANN-based DLF model for SGs. Simulation results show that the newly-proposed DLF model is able to capture the non-linearity(ies) in the history load curve, such that its accuracy is approximately 97.11%, such that the average execution time is improved by 38.50%.

As the multi-variate auto-regressive training model minimizes the forecast error to some extent, so our future directions are focused on either the improvement of this model or its replacement with a better model.

Acknowledgments: The authors would like to extend their sincere appreciation to the Visiting Professor Program at King Saud University for funding this research.

Author Contributions: All authors discussed and agreed on the idea and scientific contribution. Ashfaq Ahmad and Nadeem Javaid performed simulations and wrote simulation sections. Ashfaq Ahmad, Nadeem Javaid and Nabil Alrajeh did mathematical modeling in the manuscript. Ashfaq Ahmad, Zahoor Ali Khan, Umar Qasim and Abid Khan contributed in manuscript writing and revisions.

Conflicts of Interest: The authors declare no conflict of interest.

Bibliography

1. Unveiling the Hidden Connections between E-mobility and Smart Microgrid. Available online: http://www.zeitgeistlab.ca/doc/Unveiling_the_Hidden_Connections_between_ E-mobility_and_Smart_Microgrid.html (accessed on 20 February 2015).
2. Yan, Y.; Qian, Y.; Sharif, H.; Tipper, D. A Survey on Smart Grid Communication Infrastructures: Motivations, Requirements and Challenges. *IEEE Commun. Surv. Tutor.* **2013**, *15*, 5–20.
3. Siano, P. Demand response and smart grids: A survey. *Renew. Sustain. Energy Rev.* **2014**, *30*, 461–478.
4. Mohassel, R.R.; Fung, A.; Mohammadi, F.; Raahemifar, K. A survey on advanced metering infrastructure. *Int. J. Electr. Power Energy Syst.* **2014**, *63*, 473–484.
5. Atzeni, I.; Ordonez, L.G.; Scutari, G.; Palomar, D.P.; Fonollosa, J.R. Demand-Side Management via Distributed Energy Generation and Storage Optimization. *IEEE Trans. Smart Grid* **2013**, *4*, 866–876.
6. Adika, C.O.; Wang, L. Autonomous Appliance Scheduling for Household Energy Management. *IEEE Trans. Smart Grid* **2014**, *5*, 673–682.
7. Koutsopoulos, I.; Tassiulas, L. Optimal Control Policies for Power Demand Scheduling in the Smart Grid. *IEEE J. Sel. Areas Commun.* **2012**, *30*, 1049–1060.
8. Hermanns, H.; Wiechmann, H. Demand-Response Management for Dependable Power Grids. *Embed. Syst. Smart Appl. Energy Manag.* **2013**, *3*, 1–22.
9. Amjady, N.; Keynia, F.; Zareipour, H. Short-Term Load Forecast of Microgrids by a New Bilevel Prediction Strategy. *IEEE Trans. Smart Grid* **2010**, *1*, 286–294.
10. Gruber, J.K.; Prodanovic, M. Residential energy load profile generation using a probabilistic approach. In Proceedings of the 2012 UKSim-AMSS 6th European Modelling Symposium, Valetta, Malta, 14–16 November 2012; pp. 317–322.
11. Kou, P.; Gao, F. A sparse heteroscedastic model for the probabilistic load forecasting in energy-intensive enterprises. *Electr. Power Energy Syst.* **2014**, *55*, 144–154.
12. Amjaday, N.; Keynia, F. Day-Ahead Price Forecasting of Electricity Markets by Mutual Information Technique and Cascaded Neuro-Evolutionary Algorithm. *IEEE Trans. Power Syst.* **2009**, *24*, 306–318.
13. Yang, H.T.; Liao, J.T.; Lin, C.I. A load forecasting method for HEMS applications. In Proceedings of 2013 IEEE Grenoble PowerTech (POWERTECH), Grenoble, France, 16–20 June 2013; pp. 1–6.
14. Meidani, H.; Ghanem, R. Multiscale Markov models with random transitions for energy demand management. *Energy Build.* **2013**, *61*, 267–274.

15. Amjady, N.; Keynia, F. Electricity market price spike analysis by a hybrid data model and feature selection technique. *Electr. Power Syst. Res.* **2009**, *80*, 318–327.
16. Liu, N.; Tang, Q.; Zhang, J.; Fan, W.; Liu, J. A Hybrid Forecasting Model with Parameter Optimization for Short-term Load Forecasting of Micro-grids. *Appl. Energy* **2014**, *129*, 336–345.
17. Anderson, C.W.; Stolz, E.A.; Shamsunder, S. Multivariate autoregressive models for classification of spontaneous electroencephalographic signals during mental tasks. *IEEE Trans. Biomed. Eng.* **1998**, *45*, 277–286.
18. Engelbrecht, A.P. *Computational Intelligence: An Introduction*, 2nd ed.; John Wiley & Sons: Hoboken, NJ, USA, 2007.
19. PJM Home Page. Available online: www.pjm.com (accessed on 8 December 2015).

Improving Multi-Instance Multi-Label Learning by Extreme Learning Machine

Ying Yin, Yuhai Zhao, Chengguang Li and Bin Zhang

Abstract: Multi-instance multi-label learning is a learning framework, where every object is represented by a bag of instances and associated with multiple labels simultaneously. The existing degeneration strategy-based methods often suffer from some common drawbacks: (1) the user-specific parameter for the number of clusters may incur the effective problem; (2) SVM may bring a high computational cost when utilized as the classifier builder. In this paper, we propose an algorithm, namely multi-instance multi-label (MIML)-extreme learning machine (ELM), to address the problems. To our best knowledge, we are the first to utilize ELM in the MIML problem and to conduct the comparison of ELM and SVM on MIML. Extensive experiments have been conducted on real datasets and synthetic datasets. The results show that MIMLELM tends to achieve better generalization performance at a higher learning speed.

Reprinted from *Appl. Sci.* Cite as: Yin, Y.; Zhao, Y.; Li, C.; Zhang, B. Improving Multi-Instance Multi-Label Learning by Extreme Learning Machine. *Appl. Sci.* **2016**, *6*, 160.

1. Introduction

When utilizing machine learning to solve practical problems, we often consider an object as a feature vector. Then, we get an instance of the object. Further, associating the instance with a specific class label of the object, we obtain an example. Given a large collection of examples, the task is to get a function mapping from the instance space to the label space. We expect that the learned function can predict the labels of unseen instances correctly. However, in some applications, a real-world object is often ambiguous, which consists of multiple instances and corresponds to multiple different labels simultaneously.

For example, an image usually contains multiple patches each represented by an instance, while in image classification, such an image can belong to several classes simultaneously, e.g., an image can belong to mountains, as well as Africa [1]; another example is text categorization [1], where a document usually contains multiple sections each of which can be represented as an instance, and the document can be regarded as belonging to different categories if it were viewed from different aspects, e.g., a document can be categorized as a scientific novel, Jules Verne's writing or even books on traveling. The MIML (Multi-instance Multi-label) problem also arises in the protein function prediction task [2]. A domain is a distinct functional and structural unit of a protein. A multi-functional protein often consists of several domains, each fulfilling its own function independently. Taking a protein as an object, a domain as an instance and each biological function as a label, the protein function prediction problem exactly matches the MIML learning task.

In this context, multi-instance multi-label learning was proposed [1]. Similar to the other two multi-learning frameworks, *i.e.*, multi-instance learning (MIL) [3] and multi-label learning (MLL) [4], the MIML learning framework also results from the ambiguity in representing the real-world objects. Differently, more difficult than two other multi-learning frameworks, MIML studies the ambiguity in terms of both the input space (*i.e.*, instance space) and the output space (*i.e.*, label space), while MIL just studies the ambiguity in the input space and MLL just the ambiguity in the output space, respectively. In [1], Zhou *et al.* proposed a degeneration strategy-based framework for MIML, which consists of two phases. First, the MIML problem is degenerated into the single-instance multi-label (SIML) problem through a specific clustering process; second, the SIML problem is decomposed into a multiple independent binary classification (*i.e.*, single-instance single-label) problem using Support Vector Machine (SVM) as the classifiers builder. This two-phase framework has been successfully applied to many real-world applications and has been shown to be effective [5]. However, it could be further improved if the following drawbacks are tackled. On one hand, the clustering process in the first phase requires a user-specific parameter for the number of clusters. Unfortunately, it is often difficult to determine the correct number of clusters in advance. The incorrect number of clusters may affect the accuracy of the learning algorithm; on the other hand, SIML is degenerated into single-instance single-label learning (SISL) (*i.e.*, single instance, single label) in the second phase, as this will increase the volume of data to be handled and thus burden the classifier building. Utilizing SVM as the classifier builder in this phase may suffer from a high computational cost and require a number of parameters to be optimized.

In this paper, we propose to enhance the two-phase framework by tackling the two above issues and make the following contributions: (1) We utilize extreme learning machine (ELM) [6] instead of SVM to improve the efficiency of the two-phase framework. To our best knowledge, we are the first to utilize ELM in the MIML problem and to conduct the comparison of ELM and SVM on MIML. (2) We design a method of theoretical guarantee to determine the number of clusters automatically while incorporating it into the improved two-phase framework for effectiveness.

The remainder of this paper is organized as follows. In Section 2, we give a brief introduction to MIML and ELM. Section 3 details the improvements of the two-phase framework. Experimental analysis is given in Section 4. Finally, Section 5 concludes this paper.

2. The Preliminaries

This research is related to some previous work on MIML learning and ELM. In what follows, we briefly review some preliminaries of the two related works in Sections 2.1 and 2.2, respectively.

2.1. Multi-Instance Multi-Label Learning

In traditional supervised learning, the relationships between an object and its description and its label are always a one-to-one correspondence. That is, an object is represented by a single instance and associated with a single class label. In this sense, we refer to it as single-instance single-label learning (SISL). Formally, let X be the instance space (or say, feature space) and Y the set of class labels. The goal of SISL is to learn a function $f_{SISL}: X \rightarrow Y$ from a given dataset $\{(x_1, y_1), (x_2, y_2), \ldots, (x_m, y_m)\}$, where $x_i \in X$ is an instance and $y_i \in Y$ is the label of x_i. This formalization is prevailing and successful. However, as mentioned in Section 1, many real-world objects are complicated and ambiguous in their semantics. Representing these ambiguous objects with SISL may lose some important information and make the learning task problematic [1]. Thus, many real-world complicated objects do not fit in this framework well.

In order to deal with this problem, several multi-learning frameworks have been proposed, e.g., multi-instance learning (MIL), multi-label learning (MLL) and multi-instance multi-label Learning (MIML). MIL studies the problem where a real-world object described by a number of instances is associated with a single class label. The training set for MIL is composed of many bags each containing multiple instances. In particular, a bag is labeled positively if it contains at least one positive instance and negatively otherwise. The goal is to label unseen bags correctly. Note that although the training bags are labeled, the labels of their instances are unknown. This learning framework was formalized by Dietterich *et al.* [3] when they were investigating drug activity prediction. Formally, let X be the instance space (or say, feature space) and Y the set of class labels. The task of MIL is to learn a function $f_{MIL}: 2^X \rightarrow \{-1, +1\}$ from a given dataset $\{(X_1, y_1), (X_2, y_2), \ldots, (X_m, y_m)\}$, where $X_i \subseteq X$ is a set of instances $\{x_1^{(i)}, x_2^{(i)}, \ldots, x_{n_i}^{(i)}\}$, $x_j^{(i)} \in X (j = 1, 2, \ldots, n_i)$, and $y_i \in \{-1, +1\}$ is the label of X_i. Multi-instance learning techniques have been successfully applied to diverse applications, including image categorization [7,8], image retrieval [9,10], text categorization [11,12], web mining [13], spam detection [14], face detection [15], computer-aided medical diagnosis [16], *etc.* Differently, MLL studies the problem where a real-world object is described by one instance, but associated with a number of class labels. The goal is to learn a function $f_{MLL}: X \rightarrow 2^Y$ from a given dataset $\{(x_1, Y_1), (x_2, Y_2), \ldots, (x_m, Y_m)\}$, where $x_i \in X$ is an instance and $Y_i \subseteq Y$ a set of labels $\{y_1^{(i)}, y_2^{(i)}, \ldots, y_{l_i}^{(i)}\}$, $y_k^{(i)} \in Y (k = 1, 2, \ldots, l_i)$. The existing work of MLL falls into two major categories. One attempts to divide multi-label learning to a number of two class classification problems [17,18] or to transform it into a label ranking problem [19,20]; the other tries to exploit the correlation between the labels [21,22]. MLL has been found useful in many tasks, such as text categorization [23], scene classification [24], image and video annotation [25,26], bioinformatics [27,28] and even association rule mining [29,30].

MIML is a generalization of traditional supervised learning, multi-instance learning and multi-label learning, where a real-world object may be associated with a number of instances and a number of labels simultaneously. In some cases, transforming single-instance multi-label objects to MIML objects for learning may be beneficial. Before the explanation, we first introduce how to perform such a transformation. Let $S = \{(\mathbf{x}_1, Y_1), (\mathbf{x}_2, Y_2), \ldots, (\mathbf{x}_m, Y_m)\}$ be the dataset, where $\mathbf{x}_i \in X$ is an instance and $Y_i \subseteq Y$ a set of labels $\{y_1^{(i)}, y_2^{(i)}, \ldots, y_{l_i}^{(i)}\}$, $y_k^{(i)} \in Y (k = 1, 2, \ldots, l_i)$. We can first obtain a vector $\mathbf{v}_l$ for each class label $l \in Y$ by averaging all of the training instances of label l, i.e., $\mathbf{v}_l = \frac{1}{|S_l|} \sum_{\mathbf{x}_i \in S_l} \mathbf{x}_i$, where S_l is the set of all of the training instances $\mathbf{x}_i$ of label l. Then, each instance can be transformed into a bag, B_i, of $|Y|$ instances by computing $B_i = \{\mathbf{x}_i - \mathbf{v}_l | l \in Y\}$. As such, the single-instance multi-label dataset S is transformed into an MIML dataset $S' = \{(B_1, Y_1), (B_2, Y_2), \ldots, (B_m, Y_m)\}$. The benefits of such a transformation are intuitive. First, for an object associated with multiple class labels, if it is described by only a single instance, the information corresponding to these labels is mixed and thus difficult to learn. However, by breaking the single-instance into a number of instances, each corresponding to one label, the structure information collapsed in the single-instance representation may become easier to exploit. Second, for each label, the number of training instances can be significantly increased. Moreover, when representing the multi-label object using a set of instances, the relation between the input patterns and the semantic meanings may become more easily discoverable. In some cases, understanding why a particular object has a certain class label is even more important than simply making an accurate prediction while MIML offers a possibility for this purpose. For example, using MIML, we may discover that one object has label l_1 because it contains $instance_n$; it has label l_k because it contains $instance_i$; while the occurrence of both $instance_1$ and $instance_i$ triggers label l_j. Formally, the task of MIML is to learn a function $f_{MIML}: 2^X \to 2^Y$ from a given dataset $\{(X_1, Y_1), (X_2, Y_2), \ldots, (X_m, Y_m)\}$, where $X_i \subseteq X$ is a set of instances $\{x_1^{(i)}, x_2^{(i)}, \ldots, x_{n_i}^{(i)}\}$, $x_j^{(i)} \in X (j = 1, 2, \ldots, n_i)$ and $Y_i \subseteq Y$ is a set of labels $\{y_1^{(i)}, y_2^{(i)}, \ldots, y_{l_i}^{(i)}\}$, $y_k^{(i)} \in X(k = 1, 2, \ldots, l_i)$. Figure 1 illustrates the relationship among the four learning frameworks mentioned above.

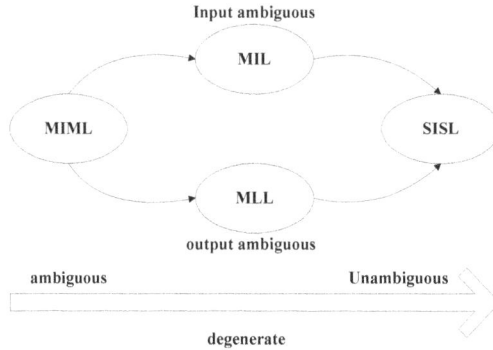

Figure 1. The relationship among these four learning frameworks.

2.2. A Brief Introduction to ELM

Extreme learning machine (ELM) is a generalized single hidden-layer feedforward network. In ELM, the hidden-layer node parameter is mathematically calculated instead of being iteratively tuned; thus, it provides good generalization performance at thousands of times faster speed than traditional popular learning algorithms for feedforward neural networks [31].

As a powerful classification model, ELM has been widely applied in many fields. For example, in [32], ELM was applied for plain text classification by using the one-against-one (OAO) and one-against-all (OAA) decomposition scheme. In [31], an ELM-based XML document classification framework was proposed to improve classification accuracy by exploiting two different voting strategies. A protein secondary prediction framework based on ELM was proposed in [33] to provide good performance at extremely high speed. The work in [34] implemented the protein-protein interaction prediction on multi-chain sets and on single-chain sets using ELM and SVM for a comparable study. In both cases, ELM tends to obtain higher recall values than SVM and shows a remarkable advantage in computational speed. The work in [35] evaluated the multi-category classification performance of ELM on three microarray datasets. The results indicate that ELM produces comparable or better classification accuracies with reduced training time and implementation complexity compared to artificial neural network methods and support vector machine methods. In [36], the use of ELM for multiresolution access of terrain height information was proposed. The optimization method-based ELM for classification was studied in [37].

ELM not only tends to reach the smallest training error, but also the smallest norm of weights [6]. Given a training set $D = \{(\mathbf{x}_i, \mathbf{t}_i) | \mathbf{x}_i \in \mathbf{R}^n, \mathbf{t}_i \in \mathbf{R}^m, i = 1, \ldots, N\}$, activation function $g(x)$ and hidden node number L, the pseudocode of ELM is given in Algorithm 1. More detailed introductions to ELM can be found in a series of published literature [6,37,38].

Algorithm 1: ELM

    Input: DB: dataset; HN: number of hidden layer nodes; AF: activation
           function
    Output: Results
 1 **for** $i = 1$ *to* L **do**
 2   randomly assign input weight w_i;
 3   randomly assign bias b_i;
 4 calculate $\mathbf{H}$;
 5 calculate $\beta = \mathbf{H}^{\dagger}\mathbf{T}$
---

3. The Proposed Approach MIMLELM

MIMLSVM is a representative two-phase MIML algorithm successfully applied in many real-world tasks [2]. It was first proposed by Zhou *et al.* in [1] and recently improved by Li *et al.*, in [5]. MIMLSVM solves the MIML problem by first degenerating it into single-instance multi-label problems through a specific clustering process and then decomposing the learning of multiple labels into a series of binary classification tasks using SVM. However, as mentioned, MIMLSVM may suffer from some drawbacks in either of the two phases. For example, in the first phase, the user-specific parameter for the number of clusters may incur the effective problem; in the second phase, utilizing SVM as the classifiers builder may bring high computational cost and require a great number of parameters to be optimized.

In this paper, we present another algorithm, namely MIMLELM, to make MIMLSVM more efficient and effective. In this proposed method: (1) We utilize ELM instead of SVM to improve the efficiency of the two-phase framework. To our best knowledge, we are the first to utilize ELM in the MIML problem and to conduct the comparison of ELM and SVM on MIML. (2) We develop a method of theoretical guarantee to determine the number of clusters automatically, so that the transformation from MIML to SIML is more effective. (3) We exploit a genetic algorithm-based ELM ensemble to further improve the prediction performance.

The MIMLELM algorithm is outlined in Algorithm 2. It consists of four major elements: (1) determination the number of clusters (Line 2); (2) transformation from MIML to SIML (Lines 3–12); (3) transformation from SIML to SISL (Lines 13–17); (4) multi-label learning based on ELM (Lines 18–19). In what follows, we will detail the four elements in Section 3.1–3.4, respectively.

Algorithm 2: The MIMLELM algorithm.

Input: DB: dataset; HN: number of hidden layer nodes; AF: activation
       function

Output: Results

1 $DB = \{(X_1, Y_1), (X_2, Y_2), \ldots, (X_m, Y_m)\}, \Gamma = X_1, X_2, \ldots, X_m;$

2 determine the number of clusters, k, using AIC;

3 randomly select k elements from Γ to initialize the k medoids
   $\{M_1, M_2, \ldots, M_k\};$

4 **repeat**

5      $\Gamma_t = \{M_t\}(t = 1, 2, \ldots, k);$

6   **foreach** $X_u \in (\Gamma - \{M_t\})$ **do**

7          $index = \arg \min_{t \in \{1,2,\ldots,k\}} d_H(X_u, M_t);$

8          $\Gamma_{index} = \Gamma_{index} \cup \{X_u\}$

9      $M_t = \arg \min\limits_{A \in \Gamma_t} \sum\limits_{B \in \Gamma_t} d_H(A, B)(t = 1, 2, \ldots, k);$

10   Transform (X_u, Y_u) into into an SIML example (z_u, Y_u), where $z_u = $
        $(d_H(X_u, M_1), d_H(X_u, M_2), \ldots, d_H(X_u, M_k));$

11 **until** M_t $(t = 1, 2, \ldots, k)$ *don't change;*

12 **foreach** z_u $(u \in \{1, 2, \ldots, m\})$ **do**

13   **foreach** $y \in Y_u$ **do**

14   decompose (z_u, Y_u) into $|Y_u|$ SISL examples

15 Train ELM_y for every class y;

16 Integrate all ELM_y's based on GA

3.1. Determination of the Number of Clusters

The primary important task for MIMLELM is to transform MIML into SIML. Unlike MIMLSVM, which performs the transformation through a clustering process with a user-specified parameter for the number of clusters, we utilize AIC [39], a model selection criterion, to automatically determine the number of clusters.

AIC is founded on information theory. It offers a relative estimation of the information lost when a given model is used to represent the process that generates the data. For any statistical model, the general form of AIC is $AIC = -2ln(L) + 2K$, where L is the maximized value of the likelihood function for the model and K is the number of parameters in the model. Given a set of candidate models, the one of the minimum AIC value is preferred [39].

Let M_k be the model of the clustering result with k clusters $C_1, C_2, \ldots, C_k$, where the number of samples in C_i is m_i. X_i denotes a random variable indicating the PD value between any pair of micro-clusters in C_i. Then, under a general assumption commonly used in the clustering community, X_i follows a Gaussian distribution with (μ_i, σ_i^2), where μ_i is the expected PD value between any pair of micro-clusters in C_i, and σ_i^2 is the corresponding variance. That is, the probability density of X_i is:

$$p(X_i) = \frac{m_i}{m} \cdot \frac{1}{\sqrt{2\pi}\sigma_i} \exp(-\frac{1}{2\sigma_i^2}(X_i - \mu_i)^2) \tag{1}$$

Let x_{i_j} $(1 \leq j \leq C_{m_i}^2)$ be an observation of X_i; the corresponding log-likelihood w.r.t the data in C_i is:

$$\ln L(C_i|\mu_i, \sigma_i) = \ln \prod_{j=1}^{C_{m_i}^2} p(X_i = x_{i_j}) = \sum_{j=1}^{C_{m_i}^2} (\ln \frac{1}{\sqrt{2\pi}\sigma_i} - \frac{1}{2\sigma_i^2}(x_{i_j} - \mu_i)^2 + \ln \frac{m_i}{m}) \tag{2}$$

Since the fact that the log-likelihood for all clusters is the sum of the log-likelihood of the individual clusters, the log-likelihood of the data w.r.t M_k is:

$$\ln L(M_k|\mu_1, \mu_2, \ldots, \mu_k, \sigma_1, \sigma_2, \ldots, \sigma_k) = \sum_{i=1}^{k} \ln L(C_i|\mu_i, \sigma_i) = \sum_{i=1}^{k} \sum_{j=1}^{C_{m_i}^2} (\ln \frac{1}{\sqrt{2\pi}\sigma_i} - \frac{1}{2\sigma_i^2}(x_{i_j} - \mu_i)^2 + \ln \frac{m_i}{m}) \tag{3}$$

Further, take the MLE (maximum likelihood estimate) of σ_i^2, i.e.:

$\widehat{\sigma_i^2} = \frac{1}{C_{m_i}^2} \sum_{j=1}^{C_{m_i}^2} (x_{i_j} - \mu_i)^2$, into Equation (3); we obtain that:

$$\ln L(M_k|\mu_1, \mu_2, \ldots, \mu_k, \sigma_1, \sigma_2, \ldots, \sigma_k) = -\frac{\sum_{i=1}^{k} C_{m_i}^2}{2} \ln(2\pi)$$

$$-\frac{\sum_{i=1}^{k} C_{m_i}^2 \ln(\widehat{\sigma_i^2})}{2} - \frac{\sum_{i=1}^{k} C_{m_i}^2}{2} + \sum_{i=1}^{k} C_{m_i}^2 \ln m_i - \ln m \sum_{i=1}^{k} C_{m_i}^2 \tag{4}$$

Finally, in our case, the number of independent parameters K is $2k$. Thus, AIC of the model M_k is:

$$AIC_{M_k} = \ln(2\pi m^2 e) \sum_{i=1}^{k} C_{m_i}^2 + \sum_{i=1}^{k} C_{m_i}^2 \ln(\widehat{\sigma_i^2}) - 2\sum_{i=1}^{k} C_{m_i}^2 \ln m_i + 4k \tag{5}$$

3.2. Transformation from MIML to SIML

With the number of clusters computed, we start to transform the MIML learning task, i.e., learning a function $f_{MIML}: 2^X \to 2^Y$, to a multi-label learning task, i.e., learning a function $f_{MLL}: Z \to 2^Y$.

Given an MIML training example, the goal of this step is to get a mapping function $z_i = \phi(X_i)$, where $\phi: 2^x \rightarrow Z$, such that for any $z_i \in Z$, $f_{MLL}(z_i) = f_{MIML}(X_i)$ if $z_i = \phi(X_i)$. As such, the proper labels of a new example X_k can be determined according to $Y_k = f_{MLL}(\phi(X_k))$. Since the proper number of clusters has been automatically determined in Section 3.1, we implement the mapping function $\phi()$ by performing the following k-medoids clustering process.

Initially, each MIML example (X_u, Y_u) $(u = 1, 2, \ldots, m)$ is collected and put into a dataset Γ (Line 1). Then, a k-medoids clustering method is performed. In this process, we first randomly select k elements from Γ to initialize the k medoids M_t $(t = 1, 2, \ldots, k)$. Note: instead of a user-specified parameter, k is an automatically-determined value by Equation (6) in Section 3.1. Since each data item in Γ, i.e., X_u, is an unlabeled multi-instance bag instead of a single instance, we employ the Hausdorff distance [40] to measure the distance between two different multi-instance bags. The Hausdorff distance is a famous metric for measuring the distance between two bags of points, which has often been used in computer vision tasks. In detail, given two bags $A = \{a_1, a_2, \ldots, a_{n_A}\}$ and $B = \{b_1, b_2, \ldots, b_{n_B}\}$, the Hausdorff distance dH between A and B is defined as:

$$d_H(A, B) = \max\{\max_{a \in A} \min_{b \in B} ||a - b||, \max_{b \in B} \min_{a \in A} ||b - a||\} \qquad (6)$$

where $||a - b||$ is used to measure the distance between the instances a and b, which takes the form of the Euclidean distance; $\max_{a \in A} \min_{b \in B} ||a - b||$ and $\max_{b \in B} \min_{a \in A} ||b - a||$ denote the maximized minimum distance of every instance in A and all instances in B and the maximized minimum distance of every instance in B and all instances in A, respectively. The Hausdorff distance-based k-medoids clustering method divides the dataset Γ into k partitions, the medoids of which are $M_1, M_2, \ldots, M_k$, respectively. With the help of these medoids, every original multi-instance example X_u can be transformed into a k-dimensional numerical vector z_u, where the i-th $(i = 1, 2, \ldots, k)$ component of z_u is the Hausdorff distance between X_u and M_i, i.e., $d_H(X_u, M_i)$. In this way, every MIML example (X_u, Y_u) $(u = 1, 2, \ldots, m)$ is transformed into an SIML example (z_u, Y_u) $(u = 1, 2, \ldots, m)$ by replacing itself with its structure information, i.e., the relationship of X_u and the k medoids. Figure 2 is an illustration of this transformation, where the dataset Γ is divided into three clusters, and thus, any MIML example X_u is represented as a three-dimensional numerical vector $z_u = (d_1, d_2, d_3)$.

After this process, we obtain the mapping function $z_i = \phi(X_i)$ such that for any $z_i \in Z$, $f_{MLL}(z_i) = f_{MIML}(X_i)$ if $z_i = \phi(X_i)$.

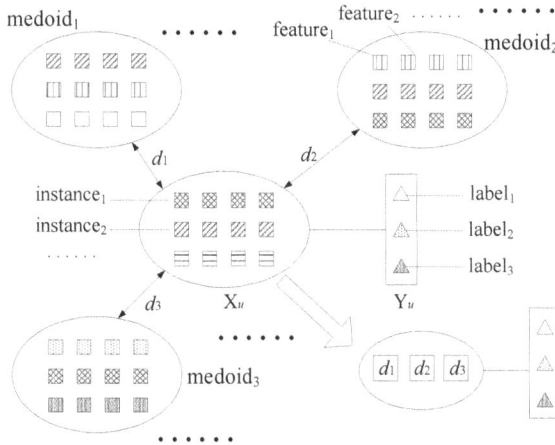

Figure 2. The process of transforming multi-instance examples into single-instance examples.

3.3. Transformation from SIML to SISL

After transforming the MIML examples (X_i, Y_i) to the SIML examples (z_i, Y_i), $i = 1, 2, \ldots m$, the SIML learning task can be further transformed into a traditional supervised learning task SISL, i.e., learning a function $f_{SISL}: Z \times Y \rightarrow \{-1, +1\}$. For this goal, we can implement the transformation from SIML to SISL in such a way that for any $y \in Y$, $f_{SISL}(z_i, y) = +1$ if $y \in Y_i$, and -1 otherwise. That is, $f_{SISL} = \{y | f_{SISL}(z_i, y) = +1\}$.

Figure 3 gives a simple illustration of this transformation. For a multi-label dataset, there are some instances that have more than one class label. It is hard for us to train the classifiers directly over the multi-label datasets. An intuitive solution to this problem is to use every multi-label data more than once when training. This is rational because every SIML example could be considered as a set of SISLs, where each SISL is of the same instance, but with a different label. Concretely, each SIML example is taken as a positive SISL example of all the classes to which it belongs. As shown in Figure 3, every circle represents an SIML example. In particular, each example in area A is of two class labels "$\bigcirc$" and "$\times$", while the other examples are of either the "$\bigcirc$" label or the "$\times$" label. According to the transformation from SIML to SISL mentioned above, an SIML example, say $(X_u, \{\bigcirc, \times\})$ in area A should be transformed into two SISL examples, $(X_{u_1}, \bigcirc)$ and $(X_{u_1}, \times)$. Consequently, when training the "$\bigcirc$" model, $(X_u, \{\bigcirc, \times\})$ is considered as $(X_{u_1}, \bigcirc)$; otherwise, it is considered as $(X_{u_1}, \times)$. In this way, the SIML examples in area A is ensured to be used as a positive example both in classes "$\bigcirc$" and "$\times$". This method can more effectively make full use of the data and make the experiment result closer to the true one.

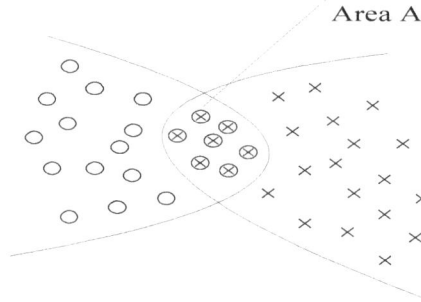

Figure 3. The example of data processing.

3.4. ELM Ensemble Based on GA

So far, we have decomposed the MIML problem into the SISL problem using SIML as the bridge. Since an MIML example is often of more than two class labels, the corresponding SISL problem should be naturally a multi-class problem.

Two commonly-used methods for multi-class classification are one-against-all (OAA) and one-against-one (OAO) [41]. For the N-class problem, OAA builds N binary classifiers, one for each class separating the class from the others. Instead, the OAO strategy involves $N(N-1)/2$ binary classifiers. Each classifier is trained to separate each pair of classes. After all $N(N-1)/2$ classifiers are trained, a voting strategy is used to make the final decision. However, a common drawback of the two strategies is that they both consider every trained classifier equally important, although the real performance may vary over different classifiers.

An ensemble classifier was proposed as an effective method to address the above problem. The output of an ensemble is a weighted average of the outputs of several classifiers, where the weights should be high for those classifiers performing well and low for those whose outputs are not reliable. However, finding the optimum weights is an optimization problem that is hard to exactly solve, especially when the objective functions do not have "nice" properties, such as continuity, differentiability, *etc*. In what follows, we utilize a genetic algorithm (GA)-based method to find the appropriate weights for each classifier.

The genetic algorithm [42] is a randomized search and optimization technique. In GA, the parameters of the search space are encoded in the form of strings called *chromosomes*. A collection of chromosomes is called a population. Initially, a random population is created. A *fitness function* is associated with each string that represents the degree of goodness of the string. Biologically-inspired operators, such as *selection*, *crossover* and *mutation*, continue for a fixed number of generations or until a termination condition is satisfied.

3.4.1. Fitness Function

Given a training instance x, the expected output of x is $d(x)$ and the actual output of the i-th individual ELM is $o_i(x)$. Moreover, let V be the validation set and $w = [w_1, w_2, \ldots, w_N]$ a possible weight assignment, $i.e.$, the chromosome of an individual in the evolving population. According to [43], the estimated generalization error of the ELM ensemble corresponding to w is:

$$E_w^V = \sum_{i=1}^{N}\sum_{j=1}^{N} w_i w_j C_{ij}^V = w^T C^V w, \tag{7}$$

where:

$$C_{ij}^V = \frac{\sum\limits_{x \in V}(f_i(x) - d(x))(f_j(x) - d(x))}{|V|} \tag{8}$$

It is obvious that E_w^V expresses the goodness of w. The smaller E_w^V is, the better w is. Thus, we use $f(w) = \frac{1}{E_w^V}$ as the fitness function.

3.4.2. Selection

During each successive generation, a certain selection method is needed to rate the fitness of each solution and preferentially select the best solution. In this paper, we use roulette wheel selection. The fitness function associated with each chromosome is used to associate a probability of selection with each individual chromosome. If f_i is the fitness of individual i in the population, the probability of i being selected is

$$p_i = \frac{f_i}{\sum\limits_{j=1}^{N} f_j} \tag{9}$$

where n is the number of individuals in the population. In this way, chromosomes with higher fitness values are less likely to be eliminated, but there is still a chance that they may be.

3.4.3. Crossover

We use the normal single point crossover. A crossover point is selected randomly between one and l (length of the chromosome). Crossover probabilities are computed as in [44]. Let f_{max} be the maximum fitness value of the current population, $\bar{f}$ be the average fitness value of the population and f' be the larger of the fitness values of the solutions to be crossed. Then, the probability of crossover, μ_c, is calculated as:

$$\mu_c = \begin{cases} k_1 \times \frac{f_{max} - f'}{f_{max} - \bar{f}}, & \text{if } f' > \bar{f}, \\ k_3, & \text{otherwise.} \end{cases} \tag{10}$$

where the values of k_1 and k_3 are kept equal to 1.0 as in [44]. Note that when $f_{max} = \overline{f}$, then $f' = f_{max}$ and μ_c will be equal to k_3. The aim behind this adaptation is to achieve a trade-off between exploration and exploitation in a different manner. The value of μ_c is increased when the better of the two chromosomes to be crossed is itself quite poor. In contrast, when it is a good solution, μ_c is low so as to reduce the likelihood of disrupting a good solution by crossover.

Mutation: Each chromosome undergoes mutation with a probability μ_m. The mutation probability is also selected adaptively for each chromosome as in [44]. That is, μ_m is given below:

$$\mu_m = \begin{cases} k_2 \times \frac{f_{max}-f}{f_{max}-\overline{f}}, & \text{if } f > \overline{f}, \\ k_4, & \text{otherwise.} \end{cases} \tag{11}$$

where the values of k_2 and k_4 are kept equal to 0.5. Each position in a chromosome is mutated with a probability μ_m in the following way. The value is replaced with a random variable drawn from a Laplacian distribution, $p(\epsilon) \propto e^{-\frac{|\epsilon-\mu|}{\delta}}$, where the scaling factor δ sets the magnitude of perturbation and μ is the value at the position to be perturbed. The scaling factor δ is chosen equal to 0.1. The old value at the position is replaced with the newly-generated value. By generating a random variable using a Laplacian distribution, there is a nonzero probability of generating any valid position from any other valid position, while the probability of generating a value near the old value is greater.

The above process of fitness computation, selection, crossover and mutation is executed for a maximum number of generations. The best chromosome seen up to the last generation provides the solution to the weighted classifier ensemble problem. Note that sum w_i should be kept during the evolving. Therefore, it is necessary to do normalization on the evolved w. Thus, we use a simple normalization scheme that replaces w_i with $w_i / \sum\limits_{i=1}^{N} w_i$ in each generation.

4. Performance Evaluation

In this section, we study the performance of the proposed MIMLELM algorithm in terms of both efficiency and effectiveness. The experiments are conducted on an HP PC (Lenovo, Shenyang, China) with 2.33 GHz Intel Core 2 CPU, 2 GB main memory running Windows 7, and all algorithms are implemented in MATLAB 2013. Both real and synthetic datasets are used in the experiments.

4.1. Datasets

Four real datasets are utilized in our experiments. The first dataset is *Image*[1], which comprises 2000 natural scene images and five classes. The percent of images of more than one class is over 22%. On average, each image is of 1.24 ± 0.46 class labels and 1.36 ± 0.54 instances; The second dataset is *Test* [22], which contains 2000 documents and seven classes. The percent of documents of multiple labels is 15%. On average, each document is of 1.15 ± 0.37 class labels and 1.64 ± 0.73 instances. The third and the fourth datasets are from two bacteria genomes, *i.e.*, *Geobacter sulfurreducens* and *Azotobacter vinelandii* [2], respectively. In the two datasets, each protein is represented as a bag of domains and labeled with a group of GO (Gene Ontology) molecular function terms. In detail, there are 397 proteins in Geobacter sulfurreducens with a total of 320 molecular function terms. The average number of instances per protein (bag) is 3.20 ± 1.21, and the average number of labels per protein is 3.14 ± 3.33. The Azotobacter vinelandii dataset has 407 proteins with a total of 320 molecular function terms. The average number of instances per protein (bag) is 3.07 ± 1.16, and the average number of labels per protein is 4.00 ± 6.97. Table 1 gives the summarized characteristics of the four datasets, where std. is the abbreviation of standard deviation.

Table 1. The information of the datasets. std.: standard deviation.

Data Set	# of Objects	# of Classes	Instances per Bag (Mean± std.)	Labels per Example (Mean ± std.)
Image	2000	5	1.36 ± 0.54	1.24 ± 0.46
Text	2000	7	1.64 ± 0.73	1.15 ± 0.37
Geobacter sulfurreducens	397	320	3.20 ± 1.21	3.14 ± 3.33
Azotobacter vinelandii	407	340	3.07 ± 1.16	4.00 ± 6.97

4.2. Evaluation Criteria

In multi-label learning, each object may have several labels simultaneously. The commonly-used evaluation criteria, such as accuracy, precision and recall, are not suitable in this case. In this paper, four popular multi-label learning evaluation criteria, *i.e.*, one-error (*OE*), coverage (*Co*), ranking loss (*RL*) and average precision (*AP*), are used to measure the performance of the proposed algorithm. Given a test dataset $S = \{(X_1, Y_1), (X_2, Y_2), \ldots, (X_p, Y_p)\}$, the four criteria are defined as below, where $h(X_i)$ returns a set of proper labels of X_i, $h(X_i, y)$ returns a real-value indicating the confidence for y to be a proper label of X_i and $rank^h(X_i, y)$ returns the rank of y derived from $h(X_i, y)$.

- $one\text{-}error_S(h) = \frac{1}{p} \sum_{i=1}^{p} \left[[\arg \max_{y \in Y} h(X_i, y)] \notin Y_i \right]$. The one-error evaluates how many times the top-ranked label is not a proper label of the object. The performance is perfect when $one\text{-}error_S(h) = 0$; the smaller the value of $one\text{-}error_S(h)$, the better the performance of h.

- $coverage_S(h) = \frac{1}{p} \sum_{i=1}^{p} \max_{y \in Y_i} rank^h(X_i, y) - 1$. The coverage evaluates how far it is needed, on the average, to go down the list of labels in order to cover all of the proper labels of the object. It is loosely related to precision at the level of perfect recall. The smaller the value of $coverage_S(h)$, the better the performance of h.

- $rloss_S(h) = \frac{1}{p} \sum_{i=1}^{p} \frac{1}{\|Y_i\|\|\overline{Y_i}\|} |\{(y_1, y_2)|h(X_i, y_1) \leq h(X_i, y_2), (y_1, y_2) \in Y_i \times \overline{Y_i}\}|$, where Y_i denotes the complementary set of Y_i in Y. The ranking loss evaluates the average fraction of label pairs that are misordered for the object. The performance is perfect when $rloss_S(h) = 0$; the smaller the value of $rloss_S(h)$, the better the performance of h.

- $avgprec_S(h) = \frac{1}{p} \sum_{i=1}^{p} \frac{1}{|Y_i|} \sum_{y \in Y_i} \frac{|\{y'|rank^h(X_i,y') \leq rank^h(X_i,y), y' \in Y_i\}|}{rank^h(X_i,y)}$. The average precision evaluates the average fraction of proper labels ranked above a particular label $y \in Y_i$. The performance is perfect when $avgprec_S(h) = 1$; the larger the value of $avgprec_S(h)$, the better the performance of h.

4.3. Effectiveness

In this set of experiments, we study the effectiveness of the proposed MIMLELM on the four real datasets. The four criteria mentioned in Section 4.2 are utilized for performance evaluation. Particularly, MIMLSVM+ [5], one of the state-of-the-art algorithms for learning with multi-instance multi-label examples, is utilized as the competitor. The MIMLSVM+ (Advanced multi-instance multi-label with support vector machine) algorithm is implemented with a Gaussian kernel, while the penalty factor cost is set from $10^{-3}, 10^{-2}, \ldots, 10^3$. The MIMLELM (multi-instance multi-label with extreme learning machine) is implemented with the number of hidden layer nodes set to be 100, 200 and 300, respectively. Specially, for a fair performance comparison, we modified MIMLSVM+ to include the automatic method for k and the genetic algorithm-based weights assignment. On each dataset, the data are randomly partitioned into a training set and a test set according to the ratio of about 1:1. The training set is used to build a predictive model, and the test set is used to evaluate its performance.

Experiments are repeated for thirty runs by using random training/test partitions, and the average results are reported in Tables 2–5, where the best performance on each criterion is highlighted in boldface, and '↓' indicates "the smaller the better", while '↑' indicates "the bigger the better". As seen from the results in Tables 2–5, MIMLSVM+ achieves better performance in terms of all cases. Applying statistical tests (nonparametric ones) to the rankings obtained for each method in the different datasets according to [45], we find that the differences are significant.

However, another important observation is that MIMLSVM+ is more sensitive to the parameter settings than MIMLELM. For example, on the Image dataset, the AP values of MIMLSVM+ vary in a wider interval $[0.3735, 0.5642]$ while those of MIMLELM vary in a narrower range $[0.4381, 0.5529]$; the C values of MIMLSVM+ vary in a wider interval $[1.1201, 2.0000]$, while those of MIMLELM vary in a narrower range $[1.5700, 2.0000]$; the OE values of MIMLSVM+ vary in a wider interval $[0.5783, 0.7969]$, while those of MIMLELM vary in a narrower range $[0.6720, 0.8400]$; and the RL values of MIMLSVM+ vary in a wider interval $[0.3511, 0.4513]$, while those of MIMLELM vary in a narrower range $[0.4109, 0.4750]$. In the other three real datasets, we have a similar observation. Moreover, we observe that in this set of experiments, MIMLELM works better when HN is set to 200.

Table 2. The effectiveness comparison on the Image data set. AP: average precision; C: coverage; OE: one-error; RL: ranking loss; MIMLSVM+: multi-instance multi-label support vector machine; MIMLELM: multi-instance multi-label-extreme learning machine.

Image		Evaluation Criterion			
		$AP\uparrow$	$C\downarrow$	$OE\downarrow$	$RL\downarrow$
MIMLSVM+	$Cost = 10^{-3}, \gamma = 2^1$	0.4999	1.2100	0.6191	0.3779
	$Cost = 10^{-2}, \gamma = 2^2$	**0.5642**	**1.1201**	**0.5783**	0.3609
	$Cost = 10^{-1}, \gamma = 2^3$	0.5142	1.1262	0.6888	**0.3511**
	$Cost = 1, \gamma = 2^1$	0.4267	1.9808	0.7391	0.3711
	$Cost = 10^1, \gamma = 2^3$	0.4705	1.9999	0.7969	0.3958
	$Cost = 10^2, \gamma = 2^5$	0.3735	1.9799	0.6809	0.4513
	$Cost = 10^3, \gamma = 2^5$	0.4541	2.0000	0.6950	0.3858
MIMLELM	$HN = 100$	0.4381	2.0000	0.8400	0.4750
	$HN = 200$	0.5529	1.7410	0.6720	0.4109
	$HN = 300$	0.4861	1.5700	0.8400	0.4376

Table 3. The effectiveness comparison on the Text dataset.

Text		Evaluation Criterion			
		$AP\uparrow$	$C\downarrow$	$OE\downarrow$	$RL\downarrow$
MIMLSVM+	$Cost = 10^{-3}, \gamma = 2^1$	0.7563	1.0295	0.3000	0.2305
	$Cost = 10^{-2}, \gamma = 2^1$	0.7675	1.0405	0.2650	0.1968
	$Cost = 10^{-1}, \gamma = 2^1$	**0.7946**	1.0445	0.2650	0.2025
	$Cost = 1, \gamma = 2^1$	0.7679	1.0145	0.2600	0.1978
	$Cost = 10^1, \gamma = 2^1$	0.7807	**1.0041**	0.2400	**0.1940**
	$Cost = 10^2, \gamma = 2^1$	0.7763	1.0450	0.2450	0.1953
	$Cost = 10^3, \gamma = 2^1$	0.7801	1.0245	**0.2350**	0.1970
MIMLELM	$HN = 100$	0.7476	1.0670	0.3540	0.2075
	$HN = 200$	0.7492	1.0928	0.3409	0.2132
	$HN = 300$	0.7554	1.0365	0.3443	0.2023

Table 4. The effectiveness comparison on the Geobacter sulfurreducens (Geob.) dataset.

Geobacter Sulfurreducens		Evaluation Criterion			
		AP ↑	C↓	OE↓	RL↓
MIMLSVM+	$Cost = 10^{-3}, \gamma = 2^5$	0.6099	1.5122	0.5583	0.2284
	$Cost = 10^{-2}, \gamma = 2^4$	0.6529	1.2439	0.5341	0.2488
	$Cost = 10^{-1}, \gamma = 2^2$	**0.6871**	**1.0488**	**0.4585**	**0.1343**
	$Cost = 1, \gamma = 2^5$	0.6755	1.0732	0.4609	0.1873
	$Cost = 10^1, \gamma = 2^3$	0.6311	1.1707	0.5097	0.1742
	$Cost = 10^2, \gamma = 2^5$	0.6733	1.1219	0.4854	0.2187
	$Cost = 10^3, \gamma = 2^1$	0.6268	1.2195	0.5097	0.2122
MIMLELM	$HN = 100$	0.6438	1.3902	0.5707	0.2151
	$HN = 200$	0.6649	1.3720	0.5390	0.2112
	$HN = 300$	0.6495	1.4025	0.5695	0.2142

Table 5. The effectiveness comparison on the Azotobacter vinelandii (Azoto.) dataset.

Azotobacter Vinelandii		Evaluation Criterion			
		AP↑	C↓	OE↓	RL↓
MIMLSVM+	$Cost = 10^{-3}, \gamma = 2^3$	0.5452	1.3171	0.6341	0.2679
	$Cost = 10^{-2}, \gamma = 2^2$	0.5652	1.0732	0.6829	0.2312
	$Cost = 10^{-1}, \gamma = 2^3$	**0.6863**	1.1707	**0.5854**	**0.1927**
	$Cost = 1, \gamma = 2^1$	0.5680	1.0488	0.6097	0.3301
	$Cost = 10^1, \gamma = 2^4$	0.6456	**1.0244**	0.6160	0.2435
	$Cost = 10^2, \gamma = 2^5$	0.5308	1.9512	0.7317	0.2150
	$Cost = 10^3, \gamma = 2^4$	0.5380	1.9756	0.6829	0.2191
MIMLELM	$HN = 100$	0.6453	1.4732	0.6414	0.2292
	$HN = 200$	0.6622	1.3610	0.6658	0.2129
	$HN = 300$	0.6574	1.4585	0.6366	0.2318

Moreover, we conduct another set of experiments to gradually evaluate the effect of each contribution in MIMLELM. That is, we first modify MIMLSVM+ to include the automatic method for k, then use ELM instead of SVM and then include the genetic algorithm-based weights assignment. The effectiveness of each option is gradually tested on four real datasets using our evaluation criteria. The results are shown in Figure 4a–d, where SVM denotes the original MIMLSVM+ [5], SVM+k denotes the modified MIMLSVM+ including the automatic method for k, ELM+k denotes the usage of ELM instead of SVM in SVM+k and ELM+k+w denotes ELM+k, further including the genetic algorithm-based weights assignment. As seen from Figure 4a–d, the options of including the automatic method for k and the genetic algorithm-based weights assignment can make the four evaluation criteria better, while the usage of ELM instead of SVM in SVM+k slightly reduces the effectiveness. Since ELM can reach a comparable effectiveness as SVM at a much faster learning speed, it is the best option to combine the three contributions in terms of both efficiency and effectiveness.

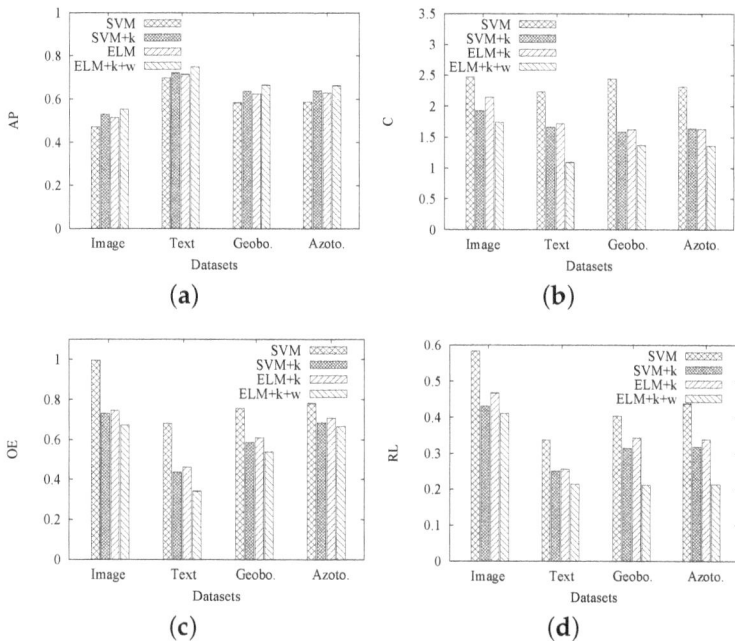

Figure 4. Gradual effectiveness evaluation of each contribution in Multi-instance Multi-label with Extreme Learning Machine (MIMLELM). (**a**) Gradual evaluation on average precision (AP); (**b**) gradual evaluation on converage (C); (**c**) gradual evaluation on one-error (OE); (**d**) gradual evaluation on ranking loss (RL).

As mentioned, we are the first to utilize ELM in the MIML problem. In this sense, it is more suitable to consider the proposed MIML-ELM as a framework addressing MIML by ELM. In other words, any better variation of ELM can be integrated into this framework to improve the effectiveness of the original one. For example, some recently-proposed methods, RELM [46], MCVELM [47], KELM [48], DropELM [49] and GEELM [50], can be integrated into this framework to improve the effectiveness of MIMLELM. In this subsection, we conducted a special set of experiments to check how the effectiveness of the proposed method could be further improved by utilizing other ELM learning processes instead of the original one. In particular, we replaced ELM exploited in our method by RELM [46], MCVELM [47], KELM [48], DropELM [49] and GEELM [50], respectively. The results of the effectiveness comparison on four different datasets are shown in Tables 6–9, respectively. As expected, the results indicates that the effectiveness of our method can be further improved by utilizing other ELM learning processes instead of the original one.

As mentioned, we are the first to utilize ELM in the MIML problem. In this sense, it is more suitable to consider the proposed MIML-ELM as a framework addressing MIML by ELM. In other words, any better variation of ELM can be integrated into this framework to improve the effectiveness of the original one. For example, some recently-proposed methods, RELM [46], MCVELM [47], KELM [48], DropELM [49] and GEELM [50], can be integrated into this framework to improve the effectiveness of MIML-ELM. In this subsection, we conducted a special set of experiments to check how the effectiveness of the proposed method could be further improved by utilizing other ELM learning processes instead of the original one. In particular, we replaced ELM exploited in our method by RELM [46], MCVELM [47], KELM [48], DropELM [49] and GEELM [50], respectively. The results of the effectiveness comparison on four different datasets are shown in Tables 6–9, respectively. As expected, the results indicate that the effectiveness of our method can be further improved by utilizing other ELM learning processes instead of the original one.

Table 6. The effectiveness comparison of Extreme Learning Machine (ELM)and its variants on the Image dataset.

Image	Evaluation Criterion			
	$AP\uparrow$	$C\downarrow$	$OE\downarrow$	$RL\downarrow$
ELM	0.5529	1.7410	0.6720	0.4109
RELM	0.7141	1.2325	0.4757	0.2909
MCVELM	0.7150	1.2239	0.4724	0.2885
KELM	0.7757	1.1346	0.4379	0.2678
DropELM	0.7814	1.1261	0.4347	0.2568
GEELM	0.7781	1.1312	0.4362	0.2667

Table 7. The effectiveness comparison of ELM and its variants on the Text dataset.

Text	Evaluation Criterion			
	$AP\uparrow$	$C\downarrow$	$OE\downarrow$	$RL\downarrow$
ELM	0.7492	1.0928	0.3409	0.2132
RELM	0.7857	1.0420	0.3251	0.2033
MCVELM	0.7959	1.0286	0.3209	0.2007
KELM	0.8019	1.0209	0.3185	0.1992
DropELM	0.8113	1.0091	0.3047	0.1906
GEELM	0.7979	1.0260	0.3198	0.2000

Table 8. The effectiveness comparison of ELM and its variants on the Geobacter sulfurreducens dataset.

Geob.	Evaluation Criterion			
	AP↑	C↓	OE↓	RL↓
ELM	0.6649	1.3720	0.5390	0.2112
RELM	0.7818	1.1668	0.4584	0.1796
MCVELM	0.7892	1.1559	0.4150	0.1626
KELM	0.8088	1.1279	0.4049	0.1586
DropELM	0.8107	1.1253	0.4020	0.1582
GEELM	0.7933	1.1499	0.4109	0.1617

Table 9. The effectiveness comparison of ELM and its variants on the Azotobacter vinelandii dataset.

Azoto.	Evaluation Criterion			
	AP↑	C↓	OE↓	RL↓
ELM	0.6622	1.3610	0.6658	0.2129
RELM	0.7928	1.1368	0.5561	0.1778
MCVELM	0.7968	1.1235	0.5533	0.1757
KELM	0.8346	1.0907	0.5283	0.1617
DropELM	0.8524	1.0679	0.5172	0.1583
GEELM	0.7997	1.1194	0.5513	0.1650

4.4. Efficiency

In this series of experiments, we study the efficiency of MIMLELM by testing its scalability. That is, each dataset is replicated different numbers of times, and then, we observe how the training time and the testing time vary with the data size increasing. Again, MIMLSVM+ is utilized as the competitor. Similarly, the MIMLSVM+ algorithm is implemented with a Gaussian kernel, while the penalty factor cost is set from $10^{-3}, 10^{-2}, \ldots, 10^3$. The MIMLELM is implemented with the number of hidden layer nodes set to be 100, 200 and 300, respectively.

The experimental results are given in Figures 5–8. As we observed, when the data size is small, the efficiency difference between MIMLSVM+ and MIMLELM is not very significant. However, as the data size increases, the superiority of MIMLELM becomes more and more significant. This case is particularly evident in terms of the testing time. In the Image dataset, the dataset is replicated 0.5–2 times with the step size set to be 0.5. When the number of copies is two, the efficiency improvement could be up to one 92.5% (from about 41.2 s down to about 21.4 s). In the Text dataset, the dataset is replicated 0.5–2 times with the step size set to be 0.5. When the number of copies is two, the efficiency improvement could be even up to 223.3% (from about 23.6 s down to about 7.3 s). In the Geobacter sulfurreducens dataset, the dataset is replicated 1–5 times with the step size set to be 1. When the number of copies is five, the efficiency improvement could be up to 82.4% (from about 3.1 s down to about 1.7 s). In the Azotobacter vinelandii dataset, the dataset is replicated 1–5 times with the step size set to be one. When the number of copies is five, the efficiency improvement could be up to 84.2% (from about 3.5 s down to about 1.9 s).

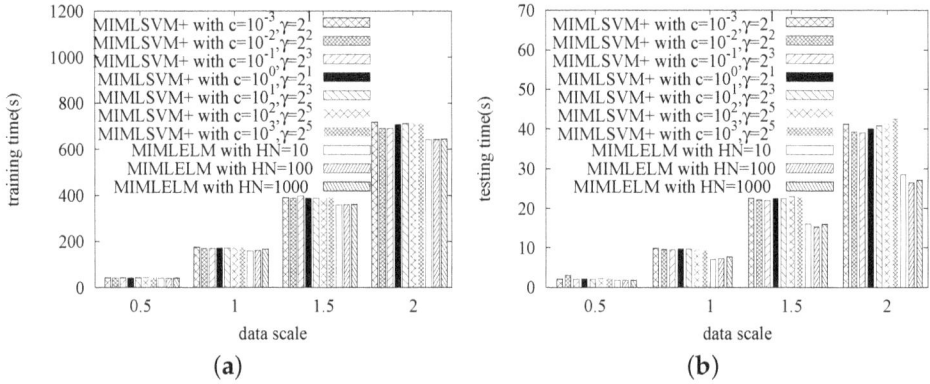

Figure 5. The efficiency comparison on the Image dataset. (**a**) The comparison of the training time; (**b**) the comparison of the testing time.

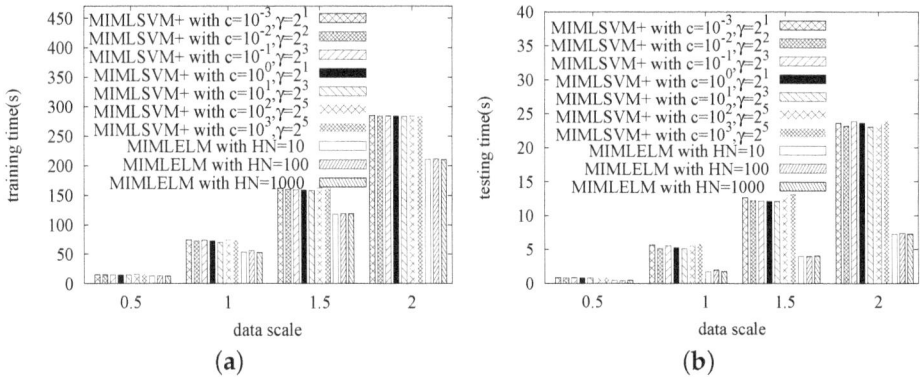

Figure 6. The efficiency comparison on the Text dataset. (**a**) The comparison of the training time; (**b**) the comparison of the testing time.

Figure 7 (a) legend:
training time(s)

MIMLSVM+ with c=10^{-3},$\gamma=2^1$
MIMLSVM+ with c=10^{-2},$\gamma=2^2$
MIMLSVM+ with c=10^{-1},$\gamma=2^3$
MIMLSVM+ with c=10^0,$\gamma=2^1$
MIMLSVM+ with c=10^1,$\gamma=2^3$
MIMLSVM+ with c=10^2,$\gamma=2^5$
MIMLSVM+ with c=10^3,$\gamma=2^5$
MIMLELM with HN=10
MIMLELM with HN=100
MIMLELM with HN=1000

data scale

(a)

Figure 7 (b) legend:
testing time(s)

MIMLSVM+ with c=10^{-3},$\gamma=2^1$
MIMLSVM+ with c=10^{-2},$\gamma=2^2$
MIMLSVM+ with c=10^{-1},$\gamma=2^3$
MIMLSVM+ with c=10^0,$\gamma=2^1$
MIMLSVM+ with c=10^1,$\gamma=2^3$
MIMLSVM+ with c=10^2,$\gamma=2^5$
MIMLSVM+ with c=10^3,$\gamma=2^5$
MIMLELM with HN=10
MIMLELM with HN=100
MIMLELM with HN=1000

data scale

(b)

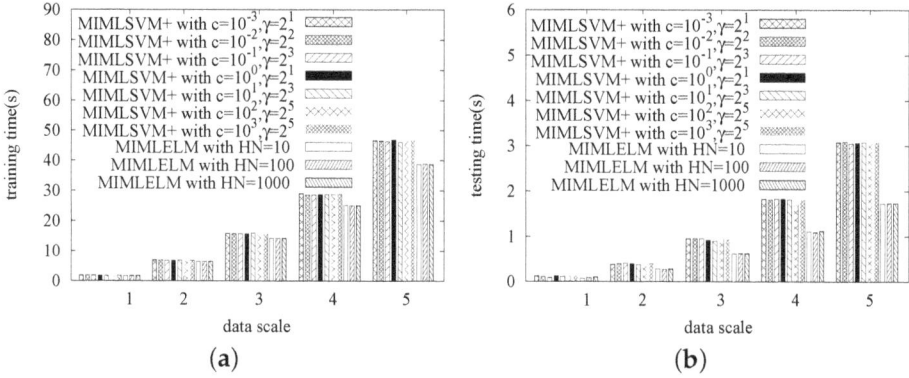

Figure 7. The efficiency comparison on the Geobacter sulfurreducens dataset. (a) The comparison of the training time; (b) the comparison of the testing time.

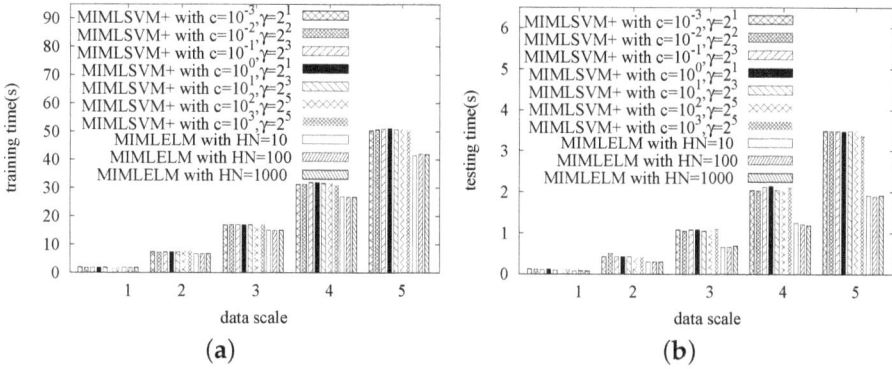

Figure 8 (a) legend:
training time(s)

MIMLSVM+ with c=10^{-3},$\gamma=2^1$
MIMLSVM+ with c=10^{-2},$\gamma=2^2$
MIMLSVM+ with c=10^{-1},$\gamma=2^3$
MIMLSVM+ with c=10^0,$\gamma=2^1$
MIMLSVM+ with c=10^1,$\gamma=2^3$
MIMLSVM+ with c=10^2,$\gamma=2^5$
MIMLSVM+ with c=10^3,$\gamma=2^5$
MIMLELM with HN=10
MIMLELM with HN=100
MIMLELM with HN=1000

data scale

(a)

Figure 8 (b) legend:
testing time(s)

MIMLSVM+ with c=10^{-3},$\gamma=2^1$
MIMLSVM+ with c=10^{-2},$\gamma=2^2$
MIMLSVM+ with c=10^{-1},$\gamma=2^3$
MIMLSVM+ with c=10^0,$\gamma=2^1$
MIMLSVM+ with c=10^1,$\gamma=2^3$
MIMLSVM+ with c=10^2,$\gamma=2^5$
MIMLSVM+ with c=10^3,$\gamma=2^5$
MIMLELM with HN=10
MIMLELM with HN=100
MIMLELM with HN=1000

data scale

(b)

Figure 8. The efficiency comparison on the Azotobacter vinelandii dataset. (a) The comparison of the training time; (b) the comparison of the testing time.

4.5. Statistical Significance of the Results

For the purpose of exploring the statistical significance of the results, we performed a nonparametric Friedman test followed by a Holm *post hoc* test, as advised by Demsar [45] to statistically compare algorithms on multiple datasets. Thus, the Friedman and the Holm test results are reported, as well.

The Friedman test [51] can be used to compare k algorithms over N datasets by ranking each algorithm on each dataset separately. The algorithm obtaining the best performance gets the rank of 1, the second best ranks 2, and so on. In case of ties, average ranks are assigned. Then, the average ranks of all algorithms on all datasets is calculated and compared. If the null hypothesis, which is all algorithms are performing equivalently, is rejected under the Friedman test statistic, *post hoc* tests, such as the Holm test [52], can be used to determine which algorithms perform statistically different. When all classifiers are compared with a control classifier and $p_1 \leq p_2 \leq \ldots \leq p_{k-1}$, Holm's step-down procedure starts with the most significant p value. If p_1 is below $\alpha/(k-1)$, the corresponding hypothesis is rejected, and we are allowed to compare p_2 to $\alpha/(k-2)$. If the second hypothesis is rejected, the test proceeds with the third, and so on. As soon as a certain null hypothesis cannot be rejected, all of the remaining hypotheses are retained, as well.

In Figure 4a–d, we have conducted a set of experiments to gradually evaluate the effect of each contribution in MIMLELM. That is, we first modify MIMLSVM+ to include the automatic method for k, then use ELM instead of SVM and then include the genetic algorithm-based weights assignment. The effectiveness of each option is gradually tested on four real datasets using four evaluation criteria. In order to further explore if the improvements are significantly different, we performed a Friedman test followed by a Holm *post hoc* test. In particular, Table 10 shows the rankings of each contribution on each dataset over criterion C. According to the rankings, we computed $\chi_F^2 = \frac{12 \times 4}{4 \times 5} \times [(4^2 + 2.25^2 + 2.75^2 + 1^2) - \frac{4 \times 5^2}{4}] = 11.1$ and $F_F = \frac{3 \times 11.1}{4 \times 3 - 11.1} = 37$. With four algorithms and four datasets, F_F is distributed according to the F distribution with $4 - 1 = 3$ and $(4-1) \times (4-1) = 9$ degrees of freedom. The critical value of $F(3,9)$ for $\alpha = 0.05$ is 3.86, so we reject the null-hypothesis. That is, the Friedman test reports a significant difference among the four methods. In what follows, we choose ELM+k+w as the control classifier and proceed with a Holm *post hoc* test. As shown in Table 11, with $SE = \sqrt{\frac{4 \times 5}{6 \times 4}} = 0.913$, the Holm procedure rejects the first hypothesis, since the corresponding p value is smaller than the adjusted α. Thus, it is statically believed that our method, *i.e.*, ELM+k+w, has a significant performance improvement of criterion C over SVM. The similar cases can be found when the tests are conducted on the other three criteria. Limited by space, we do not show them here.

In Tables 2–5, we compared the effectiveness of MIMLSVM+ and MIMLELM with different condition settings on four criteria, where, for a fair performance comparison, MIMLSVM+ is modified to include the automatic method for k and the genetic algorithm-based weights assignment as MIMLELM does. Table 12 shows the rankings of 10 classifiers on each dataset over criterion C. According to the rankings, we computed $\chi_F^2 = \frac{12 \times 4}{10 \times 11} \times [(5.5^2 + 4^2 + 3.5^2 + 3.25^2 + 3.5^2 + 6.5^2 + 6.875^2 + 8.625^2 + 7^2 + 6.25^2) - \frac{10 \times 11^2}{4}] \approx 13.43$ and $F_F = \frac{3 \times 13.43}{4 \times 9 - 13.43} \approx 1.79$. With 10 classifiers and four datasets, F_F is distributed according to the F distribution with $10 - 1 = 9$ and $(10-1) \times (4-1) = 27$ degrees of freedom. The critical value of $F(9,27)$ for $\alpha = 0.05$ is 2.25. Thus, as expected, we could not reject the null-hypothesis. That is, the Friedman test reports that there is not a significant difference among the ten methods on criterion C. This is because what we proposed in this paper is a framework. Equipped with the framework, the effectiveness of MIML can be improved further no matter whether SVM or ELM is explored. Since ELM is comparable to SVM on effectiveness [6,32,37], MIMLELM is certainly comparable to MIMLSVM+ on effectiveness. This confirms the general effectiveness of the proposed framework. Similar cases can be found when the tests are conducted on the other three criteria. Limited by space, we do not show them here.

Table 10. Friedman test of the gradual effectiveness evaluation on criterion C.

$C\downarrow$	Image	Text	Geob.	Azoto.	average rank
SVM	2.4712(4)	2.235(4)	2.439(4)	2.317(4)	4
SVM + k	1.9257(2)	1.66(2)	1.5833(2)	1.6391(3)	2.25
ELM + k	2.1451(3)	1.7198(3)	1.6247(3)	1.6285(2)	2.75
ELM + k + w	1.741(1)	1.0928(1)	1.372(1)	1.361(1)	1

Table 11. Holm test of the gradual effectiveness evaluation on criterion C.

i	Classifier	$z = (R_i - R_0)/SE$	p	$\alpha/(k-i)$
1	SVM	$(4-1)/0.913 \approx 3.286$	0.0014	0.017
2	ELM + k	$(2.75-1)/0.913 \approx 1.917$	0.0562	0.025
3	SVM + k	$(2.25-1)/0.913 \approx 1.369$	0.1706	0.05

In Figures 5–8, we studied the training time and the testing time of MIMLSVM+ and MIMLELM for the efficiency comparison, respectively. In order to further explore if the differences are significant, we performed a Friedman test followed by a Holm *post hoc* test. In particular, Table 13 shows the rankings of 10 classifiers on each dataset over training time. According to the rankings, we computed $\chi_F^2 = \frac{12 \times 4}{10 \times 11} \times [(8^2 + 5.75^2 + 6.5^2 + 7.75^2 + 5.5^2 + 7.25^2 + 8.25^2 + 1.5^2 + 2.75^2 + 1.75^2) - \frac{10 \times 11^2}{4}] \approx 26.45$ and $F_F = \frac{3 \times 26.45}{4 \times 9 - 26.45} \approx 8.31$. With ten classifiers and four datasets, F_F is distributed according to the F distribution with $10 - 1 = 9$ and $(10-1) \times (4-1) = 27$ degrees of freedom. The critical value of $F(9,27)$ for $\alpha = 0.05$ is 2.25, so we reject the null-hypothesis. That is, the Friedman test reports a significant difference among the ten methods. In what follows, we choose ELM with $HN = 100$ as the control classifier and proceed with a Holm *post hoc* test. As shown in Table 14, with $SE = \sqrt{\frac{10 \times 11}{6 \times 4}} = 2.141$, the Holm procedure rejects the hypotheses from the first to the fourth since the corresponding p-values are smaller than the adjusted α's. Thus, it is statically believed that MIMLELM with $HN = 100$ has a significant performance improvement of training over most of the MIMLSVM+ classifiers. Similarly, Table 15 shows the rankings of 10 classifiers on each dataset over testing time. According to the rankings, we computed $\chi_F^2 = \frac{12 \times 4}{10 \times 11} \times [(7.5^2 + 6.875^2 + 6.125^2 + 6.25^2 + 6.5^2 + 7.5^2 + 8.25^2 + 2.125^2 + 1.75^2 + 2.125^2) - \frac{10 \times 11^2}{4}] \approx 24.55$ and $F_F = \frac{3 \times 24.55}{4 \times 9 - 24.55} \approx 6.43$. With ten classifiers and four datasets, F_F is distributed according to the F distribution with $10 - 1 = 9$ and $(10 - 1) \times (4 - 1) = 27$ degrees of freedom. The critical value of $F(9, 27)$ for $\alpha = 0.05$ is 2.25, so we reject the null-hypothesis. That is, the Friedman test reports a significant difference among the ten methods. In what follows, we choose ELM with $HN = 200$ as the control classifier and proceed with a Holm *post hoc* test. As shown in Table 16, with $SE = \sqrt{\frac{10 \times 11}{6 \times 4}} = 2.141$, the Holm procedure rejects the hypotheses from the first to the third since the corresponding p-values are smaller than the adjusted α's. Thus, it is statically believed that MIMLELM with $HN = 200$ has a significant performance improvement of training over two of the MIMLSVM+ classifiers.

Table 12. Friedman test of the effectiveness comparison in Tables 2–5 on criterion C.

	C↓	Image	Text	Geob.	Azoto.	average rank
	$Cost = 10^{-3}, \gamma = 2^1$	1.2100(3)	1.0295(4)	1.5122(10)	1.3171(5)	5.5
	$Cost = 10^{-2}, \gamma = 2^2$	1.1201(1)	1.0405(6)	1.2439(6)	1.0732(3)	4
	$Cost = 10^{-1}, \gamma = 2^3$	1.1262(2)	1.0445(7)	1.0488(1)	1.1707(4)	3.5
MIMLSVM+	$Cost = 1, \gamma = 2^1$	1.9808(7)	1.0145(2)	1.0732(2)	1.0488(5)	3.25
	$Cost = 10^1, \gamma = 2^3$	1.9999(8)	1.0041(1)	1.1707(4)	1.0244(5)	5.5
	$Cost = 10^2, \gamma = 2^5$	1.9799(6)	1.0450(8)	1.1219(3)	1.9512(9)	6.5
	$Cost = 10^3, \gamma = 2^5$	2.0000(9.5)	1.0245(3)	1.2195(5)	1.9756(10)	6.875
	$HN = 100$	2.0000(9.5)	1.0670(9)	1.3902(8)	1.4732(8)	8.625
MIMLELM	$HN = 200$	1.7410(5)	1.0928(10)	1.3720(7)	1.3610(6)	7
	$HN = 300$	1.5700(4)	1.0365(5)	1.4025(9)	1.4585(7)	6.25

Table 13. Friedman test of the training time.

	Training Time↓	Image	Text	Geob.	Azoto.	average rank
	$Cost = 10^{-3}, \gamma = 2^1$	717.6202(10)	284.75(10)	46.582(8)	50.3727(4)	8
	$Cost = 10^{-2}, \gamma = 2^2$	690.1484(4)	283.86(7)	46.41(6)	50.6691(6)	5.75
MIMLSVM+	$Cost = 10^{-1}, \gamma = 2^3$	690.2365(5)	284.02(8)	46.27(5)	50.9343(8)	6.5
	$Cost = 1, \gamma = 2^1$	706.2458(6)	283.65(6)	46.8(9)	51.0591(10)	7.75
	$Cost = 10^1, \gamma = 2^3$	710.6634(7)	283.21(4)	46.036(4)	50.7315(7)	5.5
	$Cost = 10^2, \gamma = 2^5$	717.3216(8)	283.59(5)	46.4272(7)	50.9344(9)	7.25
	$Cost = 10^3, \gamma = 2^5$	711.5548(9)	284.47(9)	46.8312(10)	50.5936(5)	8.25
	$HN = 100$	641.3661(1)	210.55(2)	38.657(2)	41.4495(1)	1.5
MIMLELM	$HN = 200$	642.1002(2)	211.29(3)	38.922(3)	41.9643(3)	2.75
	$HN = 300$	644.2047(3)	209.84(1)	38.641(1)	41.9019(2)	1.75

In summary, the proposed framework can significantly improve the effectiveness of MIML learning. Equipped with the framework, the effectiveness of MIMLELM is comparable to that of MIMLSVM+, while the efficiency of MIMLELM is significantly better than that of MIMLSVM+.

Table 14. Holm test of the training time.

i	Classifier	$z = (R_i - R_0)/SE$	p	$\alpha/(k-i)$
1	$cost = 10^3, \gamma = 2^5$	$(8.25 - 1.5)/2.141 \approx 3.153$	0.00194	0.00556
2	$cost = 10^{-3}, \gamma = 2^1$	$(8 - 1.5)/2.141 \approx 3.036$	0.027	0.00625
3	$cost = 10^0, \gamma = 2^1$	$(7.75 - 1.5)/2.141 \approx 2.919$	0.0036	0.00714
4	$cost = 10^2, \gamma = 2^5$	$(7.25 - 1.5)/2.141 \approx 2.686$	0.0074	0.00833
5	$cost = 10^{-1}, \gamma = 2^3$	$(6.5 - 1.5)/2.141 \approx 2.335$	0.0198	0.00396
6	$cost = 10^{-2}, \gamma = 2^2$	$(5.75 - 1.5)/2.141 \approx 1.985$	0.0478	0.001195
7	$cost = 10^1, \gamma = 2^3$	$(5.5 - 1.5)/2.141 \approx 1.868$	0.0628	0.0167
8	HN = 200	$(2.75 - 1.5)/2.141 \approx 0.584$	0.562	0.025
9	HN = 300	$(1.75 - 1.5)/2.141 \approx 0.117$	0.912	0.005

Table 15. Friedman test of the testing time.

	Testing Time↓	Image	Text	Geob.	Azoto.	average rank
	$Cost = 10^{-3}, \gamma = 2^1$	41.1999(8)	23.587(7)	3.0732(7.5)	3.4944(7.5)	7.5
	$Cost = 10^{-2}, \gamma = 2^2$	39.2343(5)	23.148(5)	3.0888(10)	3.4944(7.5)	6.875
	$Cost = 10^{-1}, \gamma = 2^3$	39.1066(4)	23.834(9)	3.042(4)	3.4944(7.5)	6.125
MIMLSVM+	$Cost = 1, \gamma = 2^1$	40.0244(6)	23.615(8)	3.0576(6)	3.4788(5)	6.25
	$Cost = 10^1, \gamma = 2^3$	40.8324(7)	23.012(4)	3.0732(7.5)	3.4944(7.5)	6.5
	$Cost = 10^2, \gamma = 2^5$	41.3534(9)	23.465(6)	3.053(5)	3.4976(10)	7.5
	$Cost = 10^3, \gamma = 2^5$	742.439(10)	23.936(10)	3.0786(9)	3.3634(4)	8.25
	$HN = 100$	28.5014(3)	7.3164(1)	1.7316(2)	1.9188(2.5)	2.125
MIMLELM	$HN = 200$	26.4258(1)	7.4256(3)	1.7316(2)	1.8876(1)	1.75
	$HN = 300$	27.0154(2)	7.3457(2)	1.7316(2)	1.9188(2.5)	2.125

Table 16. Holm test of the testing time.

i	Classifier	$z = (R_i - R_0)/SE$	p	$\alpha/(k-i)$
1	$cost = 10^3, \gamma = 2^5$	$(8.25 - 1.75)/2.141 \approx 3.036$	0.0027	0.00556
2	$cost = 10^{-3}, \gamma = 2^1$	$(7.5 - 1.75)/2.141 \approx 2.686$	0.047	0.00625
3	$cost = 10^2, \gamma = 2^5$	$(7.5 - 1.75)/2.141 \approx 2.686$	0.047	0.00714
4	$cost = 10^{-2}, \gamma = 2^2$	$(6.875 - 1.75)/2.141 \approx 2.394$	0.0168	0.00833
5	$cost = 10^1, \gamma = 2^3$	$(6.5 - 1.75)/2.141 \approx 2.219$	0.0272	0.00396
6	$cost = 1, \gamma = 2^15$	$(6.25 - 1.75)/2.141 \approx 2.102$	0.0358	0.001195
7	$cost = 10^{-1}, \gamma = 2^3$	$(6.125 - 1.75)/2.141 \approx 2.043$	0.0414	0.00167
8	HN = 100	$(2.125 - 1.75)/2.141 \approx 0.175$	0.865	0.025
9	HN = 300	$(2.125 - 1.75)/2.141 \approx 0.175$	0.865	0.005

5. Conclusions

MIML is a framework for learning with complicated objects and has been proven to be effective in many applications. However, the existing two-phase MIML approaches may suffer from the effectiveness problem arising from the user-specific cluster number and the efficiency problem arising from the high computational cost. In this paper, we propose the MIMLELM approach to learn with MIML examples quickly. On the one hand, the efficiency is highly improved by integrating extreme learning machine into the MIML learning framework. To our best knowledge, we are the first to utilize ELM in the MIML problem and to conduct the comparison of ELM and SVM on MIML. On the other hand, we develop a method of theoretical guarantee to determine the number of clusters automatically and to exploit a genetic algorithm-based ELM ensemble to further improve the effectiveness.

Acknowledgments: Project supported by the National Nature Science Foundation of China (No. 61272182, 61100028, 61572117), the State Key Program of National Natural Science of China (61332014), the New Century Excellent Talents (NCET-11-0085) and the Fundamental Research Funds for the Central Universities (N150404008, N150402002, N130504001).

Author Contributions: Ying Yin and Yuhai Zhao conceived and designed the experiments; Chenguang Li performed the experiments; Ying Yin, Yuhai Zhao and Bin Zhang analyzed the data; Chenguang Li contributed reagents/materials/analysis tools; Ying Yin and Yuhai Zhao wrote the paper.

Conflicts of Interest: The authors declare no conflict of interest.

References

1. Zhou, Z.H.; Zhang, M.L. Multi-instance multi-label learning with application to scene classification. In *Advances in Neural Information Processing Systems 19*; Schagolkopf, B., Platt, J., Hoffman, T., Eds.; MIT Press: Cambridge, MA, USA, 2007; pp. 1609–1616.

2. Wu, J.; Huang, S.; Zhou, Z. Genome-wide protein function prediction through multiinstance multi-label learning. *IEEE/ACM Trans. Comput. Biol. Bioinform.* **2014**, *11*, 891–902.

3. Dietterich, T.G.; Lathrop, R.H.; Lozano-Paaerez, T. Solving the multiple instance problem with axis-parallel rectangles. *Artif. Intell.* **1997**, *89*, 31–71.

4. Schapire, R.E.; Singer, Y. Boostexter: A boosting-based system for text categorization. *Mach. Learn.* **2000**, *39*, 135–168.

5. Li, Y.; Ji, S.; Kumar, S.; Ye, J.; Zhou, Z. Drosophila gene expression pattern annotation through multi-instance multi-label learning. *IEEE/ACM Trans. Comput. Biol. Bioinform.* **2012**, *9*, 98–112.

6. Huang, G.B.; Zhu, Q.Y.; Siew, C.K. Extreme learning machine: A new learning scheme of feedforward neural networks. In Proceedings of International Joint Conference on Neural Networks (IJCNN2004). Budapest, Hungary, 25–29 July 2004; Volume 2, pp. 985–990.

7. Chen, Y.; Bi, J.; Wang, J.Z. MILES: Multiple-instance learning via embedded instance selection. *IEEE Trans. Pattern Anal. Mach. Intell.* **2006**, *28*, 1931–1947.

8. Chen, Y.; Wang, J.Z. Image categorization by learning and reasoning with regions. *J. Mach. Learn. Res.* **2004**, *5*, 913–939.

9. Yang, C.; Lozano-Paaerez, T. *Image Database Retrieval with Multiple-Instance Learning Techniques*; ICDE: San Diego, CA, USA, 2000; pp. 233–243.

10. Zhang, Q.; Goldman, S.A.; Yu, W.; Fritts, J.E. Content-based image retrieval using multipleinstance learning. In Proceedings of the Nineteenth International Conference

(ICML 2002), University of New South Wales, Sydney, Australia, 8–12 July 2002; pp. 682–689.

11. Andrews, S.; Tsochantaridis, I.; Hofmann, T. Support vector machines for multiple-instance learning. In Proceedings of the Advances in Neural Information Processing Systems 15 Neural Information Processing Systems, NIPS 2002, Vancouver, BC, Canada, 9–14 December 2002; pp. 561–568.

12. Settles, B.; Craven, M.; Ray, S. Multiple-instance active learning. In Proceedings of the Twenty-First Annual Conference on Neural Information Processing Systems, Vancouver, BC, Canada, 3–6 December 2007; pp. 1289–1296.

13. Zhou, Z.; Jiang, K.; Li, M. Multi-instance learning based web mining. *Appl. Intell.* **2005**, *22*, 135–147.

14. Jorgensen, Z.; Zhou, Y.; Inge, W.M. A multiple instance learning strategy for combating good word attacks on spam filters. *J. Mach. Learn. Res.* **2008**, *9*, 1115–1146.

15. Viola, P.A.; Platt, J.C.; Zhang, C. Multiple instance boosting for object detection. In Proceedings of the Advances in Neural Information Processing Systems 18 Neural Information Processing Systems, NIPS 2005, Vancouver, BC, Canada, 5–8 December 2005; pp. 1417–1424.

16. Fung, G.; Dundar, M.; Krishnapuram, B.; Rao, R.B. Multiple instance learning for computer aided diagnosis. In Proceedings of the Twentieth Annual Conference on Advances in Neural Information Processing Systems 19, Vancouver, BC, Canada, 4–7 December 2006; pp. 425–432.

17. Joachims, T. Text categorization with suport vector machines: Learning with many relevant features. In Proceedings of the 10th European Conference on Machine Learning, ECML-98, Chemnitz, Germany, 21–23 April 1998; pp. 137–142.

18. Yang, Y. An evaluation of statistical approaches to text categorization. *Inf. Retr.* **1999**, *1*, 69–90.

19. Elisseeff, A.; Weston, J. A kernel method for multi-labelled classification. In Proceedings of the Advances in Neural Information Processing Systems 14, NIPS 2001, Vancouver, BC, Canada, 3–8 December 2001; pp. 681–687.

20. Nigam, K.; McCallum, A.; Thrun, S.; Mitchell, T.M. Text classification from labeled and unlabeled documents using EM. *Mach. Learn.* **2000**, *39*, 103–134.

21. Liu, Y.; Jin, R.; Yang, L. Semi-supervised multi-label learning by constrained non-negative matrix factorization. In Proceedings of the Twenty-First National Conference on Artificial Intelligence and the Eighteenth Innovative Applications of Artificial Intelligence Conference, Boston, MA, USA, 16–20 July 2006; pp. 421–426.

22. Zhang, Y.; Zhou, Z. Multi-label dimensionality reduction via dependence maximization. In Proceedings of the Twenty-Third AAAI Conference on Artificial Intelligence, AAAI 2008, Chicago, IL, USA, 13–17 July 2008; pp. 1503–1505.

23. Godbole, S. Sarawagi, S. Discriminative methods for multi-labeled classification. In Proceedings of the 8th Pacific-Asia Conference on Advances in Knowledge Discovery and Data Mining, PAKDD 2004, Sydney, Australia, 26–28 May 2004; pp. 22–30.

24. Boutell, M.R.; Luo, J.; Shen, X.; Brown, C.M. Learning multi-label scene classification. *Pattern Recognit.* **2004**, *37*, 1757–1771.

25. Kang, F.; Jin, R.; Sukthankar, R. Correlated label propagation with application to multilabel learning. In Proceedings of the 2006 IEEE Computer Society Conference on Computer Vision and Pattern Recognition (CVPR 2006), New York, NY, USA, 17–22 June 2006; pp. 1719–1726.

26. Qi, G.; Hua, X.; Rui, Y.; Tang, J.; Mei, T.; Zhang, H. Correlative multi-label video annotation. In Proceedings of the 15th International Conference on Multimedia 2007, Augsburg, Germany, 24–29 September 2007; pp. 17–26.

27. Barutçuoglu, Z.; Schapire, R.E.; Troyanskaya, O.G. Hierarchical multi-label prediction of gene function. *Bioinformatics* **2006**, *22*, 830–836.

28. Brinker, K.; Fagurnkranz, J.; Hagullermeier, E. A unified model for multilabel classification and ranking. In Proceedings of the 17th European Conference on Artificial Intelligence, Including Prestigious Applications of Intelligent Systems (PAIS 2006), ECAI 2006, Riva del Garda, Italy, 29 August–1 September 2006; pp. 489–493.

29. Rak, R.; Kurgan, L.A.; Reformat, M. Multi-label associative classification of medical documents from MEDLINE. In Proceedings of the Fourth International Conference on Machine Learning and Applications, ICMLA 2005, Los Angeles, CA, USA, 15–17 December 2005.

30. Thabtah, F.A.; Cowling, P.I.; Peng, Y. MMAC: A new multi-class, multi-label associative classification approach. In Proceedings of the 4th IEEE International Conference on Data Mining (ICDM 2004), Brighton, UK, 1–4 November 2004; pp. 217–224.

31. Zhao, X.; Wang, G.; Bi, X.; Gong, P.; Zhao, Y. XML document classification based on ELM. *Neurocomputing* **2011**, *74*, 2444–2451.

32. Zhang, R.; Huang, G.B.; Sundararajan, N.; Saratchandran, P. Multi-category classification using an extreme learning machine for microarray gene expression cancer diagnosis. *IEEE/ACM Trans. Comput. Biol. Bioinform.* **2007**, *4*, 485–495.

33. Wang, G.; Zhao, Y.; Wang, D. A protein secondary structure prediction framework based on the extreme learning machine. *Neurocomputing* **2008**, *72*, 262–268.

34. Wang, D.D.; Wang, R.; Yan, H. Fast prediction of protein-protein interaction sites based on extreme learning machines. *Neurocomputing* **2014**, *128*, 258–266.

35. Zhang, R.; Huang, G.B.; Sundararajan, N.; Saratchandran, P. Multicategory classification using an extreme learning machine for microarray gene expression cancer diagnosis. *IEEE/ACM Trans. Comput. Biol. Bioinform.* **2007**, *4*, 485–495.

36. Yeu, C.-W.T.; Lim, M.-H.; Huang, G.-B.; Agarwal, A.; Ong, Y.-S. A new machine learning paradigm for terrain reconstruction. *IEEE Geosci. Remote Sens. Lett.* **2006**, *3*, 382–386.

37. Huang, G.B.; Ding, X.; Zhou, H. Optimization method based extreme learning machine for classification. *Neurocomputing* **2010**, *74*, 155–163.

38. Huang, G.B.; Zhu, Q.Y.; Siew, C.K. Extreme learning machine: Theory and applications. *Neurocomputing* **2006**, *70*, 489–501.

39. Akaike, H. A new look at the statistical model identification. *IEEE Trans. Autom. Control* **1974**, *19*, 716–723.

40. Edgar, G. *Measure, Topology, and Fractal Geometry*, 2nd ed.; Springer: New York, NY, USA, 2008.

41. Wang, Z.; Zhao, Y.; Wang, G.; Li, Y.; Wang, X. On extending extreme learning machine to non-redundant synergy pattern based graph classification. *Neurocomputing* **2015**, *149*, 330–339.

42. Goldberg, D.E. *Genetic Algorithms in Search, Optimization, and Machine Learning*; Addison-Wesley: Boston, MA, USA, 1989.

43. Zhou, Z.; Wu, J.; Jiang, Y.; Chen, S. Genetic algorithm based selective neural network ensemble. In Proceedings of the Seventeenth International Joint Conference on Artificial Intelligence, IJCAI 2001, Seattle, DC, USA, 4–10 August 2001; pp. 797–802.

44. Srinivas, M.; Patnaik, L.M. Adaptive probabilities of crossover and mutation in genetic algorithms. *IEEE Trans. Syst. Man Cybern.* **1994**, *24*, 656–667.

45. Demar, J. Statistical comparisons of classifiers over multiple data sets. *J. Mach. Learn. Res.* **2009**, *7*, 1–30.

46. Huang, G.B.; Zhou, H.; Ding, X.; Zhang R. Extreme learning machine for regression and multiclass classification. *IEEE Trans. Syst. Man Cybern. Part B Cybern.* **2012**, *42*, 513–529.

47. Iosifidis, A.; Tefas, A.; Pitas, I. Minimum class variance extreme learning machine for human action recognition. *IEEE Trans. Circuits Syst. Video Technol.* **2013**, *23*, 1968–1979.

48. Alexandros, I.; Anastastios, T.; Ioannis, P. On the kernel extreme learning machine classifier. *Pattern Recognit. Lett.* **2015**, *54*, 11–17.

49. Alexandros, I.; Anastastios, T.; Ioannis P. DropELM: Fast neural network regularization with Dropout and DropConnect. *Neurocomputing* **2015**, *162*, 57–66.

50. Alexandros, I.; Anastastios, T.; Ioannis, P. Graph embedded extreme learning machine. *IEEE Trans. Cybern.* **2016**, *46*, 311–324.

51. Friedman, M. A comparison of alternative tests of significance for the problem of *m* rankings. *Ann. Math. Stat.* **1939**, *11*, 86–92.

52. Holm S. A simple sequentially rejective multiple test procedure. *Scand. J. Stat.* **1979**, *6*, 65–70.

MDPI AG

St. Alban-Anlage 66

4052 Basel, Switzerland

Tel. +41 61 683 77 34

Fax +41 61 302 89 18

http://www.mdpi.com

Applied Sciences Editorial Office

E-mail: applsci@mdpi.com

http://www.mdpi.com/journal/applsci